Remote Pilot T

Crush the FAA Part 107 Exam on Your First Try and Skyrocket Your Career | Comprehensive Study Guide, Practice Questions, and Expert Insights

Rodney Lehman

Table of Contents

Introduction

Welcome, future remote pilot! As you delve into the pages of this study guide, you are charting a course through the expansive realm of drone aviation. In a world where the horizons of technology and possibility are ever-expanding, you stand at the forefront, ready to conquer the skies without leaving the ground. Whether it is the allure of innovative commercial applications, capturing breathtaking aerial shots, or the thrill of racing, this is where your journey begins.

Navigating the intricacies of FAA Part 107 may seem like a maze from a distance. But with this book as your guide, those complexities will resemble well-marked waypoints guiding you to your certification.

Beyond just regulations and guidelines, this guide dives deep into the nuances of the part 107 exam, offering insights and practical knowledge that will elevate your skills and increase your success rate.

Prepare for an adventure and set forth. As you study each topic, may this book spark the same excitement and curiosity that the vast skies above inspire. Here is to confidently navigate the airspace and the countless opportunities waiting on the horizon.

Clear skies ahead, aspiring remote pilot. Welcome aboard!

Why You Should Take the Exam

Successfully navigating the FAA Part 107 Exam is not merely an achievement: it is the doorway to a myriad of opportunities in commercial drone operations. With this certification, a drone pilot can tap into diverse areas that remain off-limits to uncertified operators.

This certification's unique privilege of allowing drone use for business purposes stands out. This means you can legally harness drones for ventures like aerial photography, real estate showcasing, agricultural surveillance, video making, and every other job that requires a drone. For instance, once you get certified, you could monetize your skills, offering unique perspectives at events such as weddings or sports gatherings, where aerial shots can dramatically enhance the visual experience.

The certification also grants enhanced flight privileges. With the appropriate permissions, certified pilots can fly their drones in controlled airspaces (keep reading and you will understand what they are). With the necessary waivers, there is also the potential to conduct operations during the night, thus broadening the window for potential drone activities.

Passing the part 107 exam also signifies a pilot's commitment to safety and best practices. The knowledge base required for the certification ensures a comprehensive understanding of operational safety. Remote pilots are equipped with insights into airspace navigation, weather patterns, and emergency procedures, which are invaluable in risk assessment and management during drone flights.

Certified drone operators can also participate in specialized missions. Their expertise can be invaluable in search and rescue missions, offering an aerial advantage to cover vast terrains swiftly. Furthermore, sectors like infrastructure rely on these pilots for inspecting critical structures such as bridges, towers, and pipelines.

And, most importantly, in the fast-evolving drone industry, staying updated is crucial. With the FAA Part 107 certification, pilots are more likely to be coordinated with the latest regulations, technology, and best practices. The legal peace of mind it offers is invaluable. Certified pilots can operate confidently, knowing they are within the FAA's legal framework and in a better position to obtain insurance, safeguarding against potential liabilities.

All this to say that by deciding to take the test, you are on the right path to turning your passion for drones into something bigger. Path yourself on the back for your commitment and let us get started!

How to Register for the Exam

Let us start from the basics. You can find all the information you need to register for the exam here. For most it might seem unnecessary to go over this process but trust me: I have seen countless aspiring remote pilots having trouble with the registration.

1. Prerequisites:

Before registering for the exam, ensure that you:

- Are at least 16 years old.
- are in a state of health and mind necessary for safely flying a drone.

2. Registration Process:

- Visit the FAA's testing service website (www.faa.gov).
- Create a new account or sign in with an existing one..
- From the list of available exams, select the "Part 107 Remote Pilot Exam."
- Select a convenient date and testing center location.
- Pay the registration cost, which is $175 as of this writing. Ensure to check for the most recent fee on the FAA's official site.

3. What to Bring to the Exam:

- A government-issued photo ID.
- Your FAA confirmation email or printout of your exam registration.
- A basic calculator if you prefer, though most testing centers will provide one.
- A magnifying glass. This is useful for the sectional charts related questions.

4. The Exam Structure:

- There are 60 multiple-choice questions on the test.
- The test will last two hours for you to finish.
- A passing score is 70% or better. This means you must answer correctly to at least 42 questions.

5. Receiving Your Results:

- Immediately after completion, you will receive a preliminary score.
- The official results, along with feedback on areas of weakness (if applicable), will be available on the FAA's website within 48-72 hours.

6. Once you have passed the exam:

- Complete the FAA Airman Certificate and/or Rating Application to receive your remote pilot certificate.
- Within a few of weeks, anticipate receiving your official certificate in the mail.

- Once you receive the certificate, you start operating drones for commercial purpose.

We delve further into the subjects you need to learn to ace the test in the ensuing chapters. Since this exam is not exclusive to drone pilots, you will find a lot of aviation information that is irrelevant to you in the day-to-day practice. However, you will get asked about them in the exam. They are included in our study guide because of this.

Chapter 1 – Airspace Categories, Requirements for Operations, and Flight Restrictions

Airspace is primarily divided into two main categories: regulatory and nonregulatory. These categories further branch into four distinct types: controlled, uncontrolled, particular use, and other airspace. The criteria for designating airspace types and categories are based on factors such as the volume and intricacy of aircraft operations, the specific operations taking place within that airspace, requisite safety standards, and national and public significance considerations. This chapter explains the characteristics of every airspace type.

Controlled Airspace

Controlled airspace encompasses various classifications and specifically delineated areas where the provision of air traffic control services aligns with the designated airspace classification. For remote pilots, the controlled airspace segments that are of relevance include:

- Class B
- Class C
- Class D
- Class E

Let us go at them one by one.

Class B Airspace

Class B airspace extends from the ground up to 10,000 feet mean sea level and is positioned around the country's most bustling airports, gauged by the volume of flight operations or the number of passengers boarding. Each Class B airspace has a unique configuration comprising a surface region and multiple layers. Some of these airspaces can be visualized as inverted wedding cakes. They are structured to encompass all known instrument procedures from when an aircraft joins the airspace. For remote pilots, securing authorization from air traffic control before navigating Class B airspace is imperative.

Class C Airspace

Class C airspace begins at surface level and rises to 4,000 feet above the airport's elevation, measured in mean sea level. This airspace includes airports equipped with an active control tower, supported by a radar approach control, and having a specific volume of instrument flight rules (IFR) activities or passenger boardings. Each Class C zone is custom designed. It features a surface zone with a radius of five nautical miles and an outer area that spans from 1,200 feet to 4,000 feet above the airport's elevation with a ten nautical miles radius. Before venturing into Class C airspace, remote pilots must obtain air traffic control authorization.

Class D Airspace

Measuring at mean sea level, Class D airspace extends from the ground to 2,500 feet above the airport's elevation. It is associated with airports that operate with a control tower. Each Class D zone has a bespoke configuration. When there are published instrument procedures, the airspace is typically structured to enclose those procedures. Extensions for instrument approach procedures (IAPs) may fall under the classification of Class D airspace or Class E airspace. Before entering Class D airspace, remote pilots must obtain authorization from air traffic control.

Class E Airspace

Class E airspace represents the controlled regions not categorized under Class B, C, or D., a vast portion of the airspace across the United States is designated as Class E, ensuring ample space to manage and separate aircraft safely during instrument flight rules operations.

Charts such as sectional ones pinpoint all Class E airspace locations with bases below 14,500 feet MSL. In regions where the Class E base is not depicted on the charts, the airspace begins at 14,500 feet MSL. Typically, the starting point of Class E airspace is 1,200 feet above the ground level (AGL). However, this base could be at the surface or 700 feet AGL in several locations. There are instances where Class E airspace commences at a mean sea level altitude displayed on the charts, differing from AGL measurements. Class E airspace rises to 18,000 feet MSL but does not include that height, marking the beginning of Class A airspace. Any airspace exceeding light level (FL) 600 falls under Class E.

Federal Airways is shown as blue lines in Class E airspace on sectional charts. These airways commence at 1,200 feet AGL and extend up to, but do not encompass, 18,000 feet MSL.

Remote pilots typically are not required to seek air traffic control authorization when navigating Class E airspace.

Uncontrolled Airspace (Class G)

Class G airspace, often called uncontrolled airspace, represents the sections of the airspace not categorized under other classes. It inherently carries the label of being "uncontrolled." Class G airspace starts from the ground level and stretches upward until it reaches the beginning of the immediately above Class E airspace. Remote pilots are not mandated to obtain air traffic control clearance when flying within this type of airspace.

Special Use Airspace

Also known as Special Area of Operation (SAO), this type of airspace is a designated region where specific activities are restricted or where particular limitations might be imposed on aircraft not involved in those selected activities. Such airspace can sometimes pose constraints on the concurrent use of the airspace. Information about special-use airspace illustrated on instrument charts encompasses the area's name or number, effective altitude, operational times and weather conditions, the overseeing agency, and the location on the chart panel. This data can be found on one of the end panels for those referencing National Aeronautical Charting Group route charts. The most common types of Special Use Airspaces are:

- Prohibited zones
- Restricted zones
- Warning zones
- Military Operation Areas (MOAs)
- Alert zones
- Controlled Firing Areas (CFAs)

Let us dive deeper into each one of them.

Prohibited Areas

Prohibited zones are specific sectors of the airspace where aircraft movement is strictly disallowed. These zones are set up due to security considerations or other factors crucial to national interests. Aeronautical charts and the Federal Register both contain information on these zones. These regions are marked with a "P" followed by a sequence number (like P-4808). An example of such a zone might be the vicinity around Area 51 in Nevada.

Restricted Areas

Restricted zones house airspace in which, while aircraft flight is not entirely forbidden, it is subjected to constraints. Such zones are designated due to the specific nature of the activities occurring within them, or there might be limitations placed on flying activities not associated with those special operations—or both reasons. Indicators of restricted areas highlight the potential presence of atypical and sometimes unseen dangers for aircraft, such as artillery launches, air-to-air shooting drills, or guided missile tests. Unauthorized entry into these zones without consent from the overseeing or regulatory agency can pose severe threats to the aircraft.

When a restricted zone is inactive and handed over to the FAA, the air traffic control allows aircraft to navigate this airspace without requiring a dedicated clearance. However, if the restricted area is active and remains outside of the FAA's authority, the air traffic control ensures they provide a route clearance that guarantees the aircraft sidesteps the limited zone.

Such restricted zones are symbolized on charts with an "R" prefix, followed by a specific number, like R-2916. This number represents a limited area around Eglin Air Force Base in Florida.

Warning Areas

Warning areas have characteristics like restricted areas. However, the distinction lies in that the U.S. government does not possess exclusive authority over this airspace. These areas stretch from 3 NM beyond the U.S. coastline and may house activities potentially dangerous to aircraft not involved in those operations. Their primary role is to signal to pilots who need to participate about the possible risks. These zones can span domestic or international waters or, occasionally, both. These regions are marked on charts with a "W" prefix, followed by a specific identifier, such as W-285. This description, for instance, applies to a region off the coast of Southern California.

Military Operation Areas

Military Operation Areas are characterized by specific vertical and horizontal boundaries designated to segregate certain military training exercises from instrument flight rules aircraft movements. When an MOA is active, instrument flight rules aircraft not involved in the activities might be permitted to traverse the MOA, granted that air traffic control can ensure appropriate instrument flight rules separation. In cases where this separation cannot be provided, air traffic control will redirect or limit the movement of the nonparticipating instrument flight rules aircraft. These areas are illustrated on sectional charts, visual flight rules (VFR) terminal region, and low altitude en route charts. Unlike other zones, they are not assigned a unique number but are labeled by name, such as "Ironwood MOA." Additional details about the MOA, including operational hours, affected altitudes, and the managing entity, can be found on the reverse side of the sectional charts.

Alert Areas

Alert zones are marked on aviation maps with an "A" prefix, followed by a specific sequence. For instance, A-315 points to a zone near Memphis, Tennessee. These zones serve as an advisory to pilots not involved in local operations, signaling areas that might experience dense pilot training or atypical aerial movements. While traversing alert zones, aviators should be particularly vigilant. It is crucial to understand that all maneuvers within an alert zone adhere to standard regulations without exceptions. Whether it's aircraft involved in the local operations or those just passing through, every pilot shares equal responsibility for averting collisions.

Controlled Firing Areas (CFAs)

Military Operation Areas host operations that could pose risks to aircraft not involved in those activities if not managed under stringent conditions. The mandatory pause in activities sets Controlled Firing Areas apart from other airspace categories. These operations cease when monitoring mechanisms like a spotter plane, radar, or land-based observation posts signal the potential approach of an aircraft. They are not specifically marked on charts because they do not require aircraft not involved in the activities to alter their routes.

Other Airspace Areas

Local Airport Advisory (LAA)

The Local Airport Advisory is a guidance service presented by Flight Service stations positioned at the airport. It operates via a specific ground-to-air channel or the tower channel during its non-operational hours. The Local Airport Advisory offers insights about the local airport, automated weather updates with vocal transmissions, and uninterrupted data display from systems like the Automated Surface Observing System or the Automated Weather Observing Station.

For instance, when the central tower halts operations post-evening at the Santa Monica Airport in California, the Flight Service facility can provide pilots Local Airport Advisory, offering them essential local insights and current weather conditions via automated systems.

Military Training Routes (MTRs)

Military Training Routes are dedicated routes that military aircraft utilize to hone their skills in strategic aviation. Typically, these pathways are set below 10,000 feet MSL, catering to operations with speeds surpassing 250 knots. While most of the path remains at these lower altitudes, specific segments might stretch higher to ensure route cohesion. The routes are categorized as instrument flight rules (designated as 'IF') and visual flight rules (denoted as 'VF'), succeeded by a numerical identifier.

Military Training Routes that remain entirely below 1,500 feet AGL use a four-digit code (like IF3012, VF3013). Conversely, Military Training Routes with parts that rise above 1,500 feet AGL use a three-digit system (like IF312, VF313). Instrument flight rules low-level routing charts display all IF routes and VF routes that facilitate operations over 1,500 feet AGL. IF routes follow instrument flight rules protocols, irrespective of the prevailing weather conditions. Pilots can refer to visual flight rules regional charts for military activity insights, such as IF, VF, Tactical Aviation Zone (TAZ), limited zones, precaution zones, and notification zones.

Temporary Flight Restrictions (TFR)

A Notice to Airmen (NOTAM) from the flight data center designates a Temporary Flight Restriction. Such a NOTAM typically commences with the words "AIRSPACE LIMITATIONS," detailing the specific location of the limitation, the period it is active, its statute-mile radius, and the impacted altitudes. Additionally, the NOTAM offers contact details for the coordinating FAA facility, the rationale for the Temporary Flight Restriction, and any other pertinent information. The conclusion of this chapter contains more information about NOTAMs.

Temporary Flight Restrictions are established for reasons including but not limited to:

- Shielding airborne or ground-based individuals and assets from impending or present risks.
- Making sure that aircraft used in disaster relief missions are safe.
- Mitigating the risk of overcrowded skies above events or incidents garnering significant public attention.
- Safeguarding national disaster zones for compassionate reasons, especially in Hawaii.
- Ensuring the safety of high-profile personalities, like the President or Vice President.
- Facilitating secure conditions for space agency missions.

In the aftermath of the September 11, 2001, attacks, Temporary Flight Restrictions implementations have become notably more frequent. Several instances have been reported where aircraft mistakenly entered Temporary Flight Restrictions zones, leading to security assessments and the suspension of pilot certificates. Pilots are thus mandated to remain updated on any Temporary Flight Restrictions within their anticipated flight region. The FAA's official Temporary Flight Restrictions site, www.tfr.faa.gov, serves as a reliable resource for this.

Published Visual Flight Rules Routes (VFR)

These pathways are crafted for visual flight rules flights navigating around, below, or through intricate airspace structures. Designations such as VFR flyway, VFR passageway, Class B airspace VFR transition channel, and metropolitan VFR pathway are used to label these routes. Typically, these paths can be located on visual flight rules metropolitan area navigation charts.

Real-life Examples:

- In Los Angeles, the "Hollywood Park Route" lets pilots transition north and south through the LAX Class B airspace.
- In the complex airspace of New York City, the "Hudson River Corridor" allows pilots to transition up and down the Hudson River, providing a close view of Manhattan's skyline.
- The "Shoreline Route" in San Francisco provides pilots a path along the coast, keeping them clear of the busy SFO Class B airspace.

Terminal Radar Service Areas (TRSAs)

Terminal Radar Service Areas provide zones where pilots, upon opting in, gain access to enhanced radar services. This service aims to facilitate separation between all instrument flight rules movements and visual flight rules aircraft that choose to participate. Within a Terminal Radar Service Area, the main airport transitions into Class D airspace. The bulk of the Terminal Radar Service Area encompasses other designated airspace, typically Class E airspace that starts at 700 or 1,200 feet and is geared for en route or terminal environment transitions. Terminal Radar Service Areas are illustrated on visual flight rules sectional charts and metropolitan area maps with defined black lines, specifying altitudes for distinct sections. The Class D segment is marked using a segmented blue line. While Terminal Radar Service Area services are optional, visual flight rules pilots are strongly recommended to reach out to radar approach control to leverage these services.

Real-life Examples:

- The Roanoke TRSA in Virginia provides radar services for flights near the Roanoke-Blacksburg Regional Airport.
- The Terre Haute TRSA in Indiana is established around the Terre Haute International Airport-Hulman Field, where radar services are available for both visual flight rules and instrument flight rules flights.
- Improved radar services are provided by Peoria TRSA in Illinois surrounding General Wayne A. Downing Peoria International Airport.

National Security Areas (NSAs)

National Security Areas define specific airspace volumes established where there is a need to enhance security around ground installations. Under certain circumstances, flight activities within this area type might face restrictions as per the guidelines of Title 14 of the Code of Federal Regulations, part 99. Such constraints are communicated using Notices to Airmen. It is a courtesy request for pilots to avoid these marked areas.

Real-life Examples:

- NSA around Fort Knox in Kentucky is a zone where flights are frequently advised to maintain caution due to heightened security associated with the U.S. Bullion Depository located there.
- The United States' Aberdeen Proving Ground is located in Maryland. Army tests its munitions, has an NSA around it due to the sensitive nature of its operations.
- Kings Bay Naval Submarine Base in Georgia is another area of increased security where submarines, especially those carrying ballistic missiles, are based. An NSA around this area ensures increased protection and security.

Air Traffic Control and the Unified Airspace Framework

The main goal of the Air Traffic Control structure is to avoid potential collisions between aircraft and streamline the movement of airborne traffic. Beyond this primary role, air traffic control possesses the capability, albeit within specific constraints, to offer added services, including traffic density, radio frequency crowding, radar quality, the workload of the controller, and other priority tasks. There may also be instances where monitoring and detecting specific scenarios is physically challenging. While these limitations are acknowledged, controllers are obliged to offer these supplementary services whenever circumstances allow. Such assistance is not discretionary but is mandatory whenever the operational context permits. air traffic control services follow standard procedures unless otherwise stipulated in formal agreements, FAA guidelines, or military directives.

Piloting Guidelines and Necessary Provisions

Flight safety is paramount for every pilot, and the onus of maneuvering an aircraft warrants utmost diligence. Under the vigilant supervision of the FAA, the airborne traffic system ensures an unmatched degree of safety and operational efficiency. Pilots operate under a set of regulations that have bolstered the reputation of the U.S., establishing it as the global leader in aviation safety standards.

Every aircraft soaring in the current Unified Airspace Framework adheres to the Code of Federal Regulation (CFR) protocols that dictate its certification and upkeep. Integral to safe operations are meticulous pre-flight preparations, adept aeronautical decision-making, and proactive risk mitigation. As we will see in the following chapters, aeronautical decision-making employs a holistic strategy for gauging risks and managing aviation-related stressors. It underscores how intrinsic attitudes can shape decisions and suggests ways to recalibrate such perspectives to bolster flight safety.

Notices to Airmen (NOTAMs)

Notices to Airmen, commonly called NOTAMs, serve as essential alerts providing timely aeronautical details. These details are either of a temporary nature or were unpredictably available in time for regular publication in aeronautical charts or related operational manuals. Such critical information is rapidly shared through the National Airmen Bulletin System. NOTAMs encompass up-to-the-minute advisories vital for ensuring flight safety and offer additional details that influence other operational guides. They are dispatched for a plethora of reasons, some of which include:

- Potential hazards, such as aviation exhibitions, skydiving activities, kite events, and space launches.
- Official travels by dignitaries or global leaders.
- Malfunctioning illumination on high-rise structures.
- Temporary establishment of hindrances around airports.
- Migratory patterns of bird swarms across the sky, leading to a specific NOTAM known as a BIRDTAM.

Historically, there have been real-life NOTAMs addressing situations like:

- The annual Albuquerque International Balloon Fiesta, warning pilots of the numerous hot air balloons in the area.
- Temporary flight restrictions during significant political events, like the G7 Summit.
- Bird migration patterns, especially during specific seasons around certain airports to mitigate the risk of bird strikes.

Notices to Airmen can be accessed in printed formats via subscriptions from the Document Oversight Office or digitally through platforms like PilotWeb, which offers the latest NOTAM insights. Various online portals also offer localized airport NOTAM details.

Chapter 2 - Aviation Weather Sources and Reports

Providing weather information is a collaborative effort involving the National Weather Service, the Federal Aviation Administration, the Department of Defense, and various other aviation entities and experts. Though it is a challenge to guarantee absolute accuracy in weather forecasting, contemporary meteorology, backed by scientific analysis and advanced computer simulations, offers increasingly precise insights into weather dynamics and shifts. A sophisticated network of weather agencies, governmental bodies, and independent weather enthusiasts ensures that aviators and related professionals are equipped with timely and relevant weather data. This wealth of information aids pilots in making knowledgeable decisions concerning weather implications for their flights, both before taking off and while airborne.

Surface Aviation Weather Observations

Surface aviation weather observations offer a detailed account of current weather conditions at specific terrestrial locations throughout the United States.

Each surface report outlines local weather particulars for a designated airport. Details encompass the report's nature, station code, recording date and time, any necessary modifiers, prevailing winds, visibility parameters, runway visual range, observed weather events, sky status, temperature and dew point metrics, altimeter measurements, and pertinent notes. The data for these reports can be sourced from human observers, automated stations, or automated systems supplemented by human input. Regardless of its origin, every surface report offers essential insights regarding specific airports nationwide.

Aviation Weather Reports

Weather reports in the realm of aviation are crafted to offer precise current weather details. Each report provides real-time data that is updated at various intervals. METARs and TAFs are common examples of these reports. For accessing a weather report, one can navigate to www.aviationweather.gov.

Standard Meteorological Report (METAR)

METARs provide current surface weather data presented in a globally accepted format. Typically, they are distributed every hour unless there is a notable change in weather conditions, prompting a special METAR that can be released anytime between standard METAR releases.

Here is a practical example:
METAR KLAX 141930Z AUTO 15018G24KT 2SM -RA FG SCT006 OVC010CB 16/15 A2985 RMK PRESRR

Let us break it down. A usual METAR report offers the subsequent details in order:

- Report Type—If it is a simple METAR or a special METAR (SPECI)
- Station Code—This is a four-character code as set by the International Civil Aviation Organization. In the U.S., a distinct three-letter code is prefixed by the letter "K." For instance, Los Angeles International Airport has the code "KLAX." Elsewhere globally, the first two characters of the ICAO code indicate the region or country.
- Recording Date and Time—Displayed as a six-digit figure (141930Z), the first two represent the date while the last four are the UTC time. "Z" signifies Zulu time (UTC).
- Modifier—Indicates if the METAR/SPECI is from an automated source or corrected. "AUTO" means it is automated. When "COR" appears, it denotes a corrected report.
- Wind Details—Typically a five-digit figure (15018KT) where the initial three digits indicate the wind direction in tens of degrees. The next two numbers represent wind speed in knots. Wind gusts, variability, and extreme directions have specific representations.
- Visibility Metrics—Visibility is depicted in statute miles, like "2SM." Occasionally, the runway visual range (RVR) is shared post the prevailing visibility. RVR reflects the pilot's visibility on a moving aircraft's runway.
- Weather Conditions—These divides into qualifiers and observed phenomena (-RA FG). Intensity, proximity, and descriptors come first. This is followed by precipitation types, obscurations, and other relevant phenomena.
- Sky Status—Reports in sequence of amount, altitude, and type or indefinite ceiling (SCT006 OVC010CB). Hundreds of feet above sea level is where clouds begin to form. Specific cloud types might also be reported.
- Temperature and Dew Point Data—Given in Celsius. If below zero, it is prefixed with "M" for minus (16/15).
- Barometric Pressure Reading—Reported as inches of mercury in a four-figure group (A2985). Potentially included in the statements are pressure trends.
- Zulu Time (UTC)—Used universally in aviation to synchronize global operations.
- Remarks—Starting with "RMK," this section might include additional weather specifics or comments on the METAR.

Adding it all together, this is the correct interpretation: Routine Standard Meteorological Report for Los Angeles International Airport on the 14th day at 1930Z from an automated source. Winds are coming from 150° at 18 knots, gusting to 24. Visibility stands at 2 statute miles. Light rain and fog are present. There's scattered cloud cover at 600 feet and an overcast layer at 1,000 feet with cumulonimbus clouds. Temperature is 16°C with a dew point of 15°C. The barometric pressure is rapidly increasing at 29.85 "Hg.

Exercises:

Now, to put your skills to the test, here are three METAR reports for you to interpret on your own. Use the breakdown structure above to help guide your interpretation. After you've finished your interpretations, you can review your responses in this book's "ANSWERS" section.

1. METAR KJFK 221152Z 28016KT 10SM FEW035 SCT070 BKN200 03/M03 A3005 RMK AO2 SLP174 T00281028 10028 21044 53015

2. METAR EGLL 221150Z 07012KT 040V100 9000 -DZ FEW008 SCT015 BKN040 05/03 Q1018 NOSIG

3. METAR YSSY 221200Z 04008KT 9999 FEW030 SCT048 22/16 Q1016

Terminal Aerodrome Forecast (TAF)

A Terminal Aerodrome Forecast is formulated to cater to the atmospheric conditions within a five-statute mile radius of an airport, encompassing significant airports. The validity of each Terminal Aerodrome Forecast stretches across 24 to 30 hours and undergoes revisions four times daily at intervals of 0000Z, 0600Z, 1200Z, and 1800Z. TAFs employ the same set of indicators and shorthand as Standard Meteorological Reports. When preparing for a flight, these forecasts can be an invaluable tool for remote pilots.

Here is a practical example:
TAF KJFK 121400Z 1214/1314 TEMPO 1214/1216 4SM HZ FM1600 17010G20KT P6SM SCT035 BKN200 FM122300 15010KT P6SM BKN070 OVC130 PROB30 1300/1303 4SM TSRA BKN025CB FM130500 1505KT P6SM SCT030 OVC070 TEMPO 1305/1309 4SM TSRA OVC025CB

Sequentially, the Terminal Aerodrome Forecast provides:

- Report Nature: Distinguishing between a regular forecast (TAF) or a modified one (TAF AMD).
- ICAO Station Code: This remains consistent with Standard Meteorological Reports.
- Origination Date & Time: Demonstrated as a six-digit code, e.g., 071530Z, where the initial two digits represent the date and the latter four reflect the time in UTC, marked by the following 'Z'.
- Forecast Validity Window: Immediately after the origination timestamp, this period, e.g., 0716/0816, displays the forecast's starting and concluding periods. A 24 or 30-hour forecast, dispensed at 0000, 0600, 1200, and 1800Z, starts with the day of issuance (e.g., 07) followed by the forecast's beginning hour in UTC (16). The ensuing two digits display the concluding day (08) and hour (16). A start or end at midnight UTC is denoted as 00 or 24, respectively.

- Predicted Winds: Denoted by a five-digit group, e.g., 17012KT, with the leading three depicting wind direction (relative to true north) and the trailing two signifying wind speed in knots.
- Visibility Predictions: Displayed in statute miles, either as whole numbers or fractions. Forecasts surpassing six miles are indicated by "P6SM."
- Anticipated Weather Events: These are coded identically to Standard Meteorological Reports.
- Sky Status Forecast: Also, like METAR, but only cumulonimbus clouds (CB) are considered.
- Forecast Shifts: Major weather changes expected during the Terminal Aerodrome Forecast period, including the projected conditions and their duration, can be found here. Keywords "from (FM)" and "temporary (TEMPO)" might appear. "FM" pinpoints quick, noteworthy changes, while "TEMPO" alludes to short-lived weather variations.
- PROB30: Describes the likelihood percentage of thunderstorms and rain within the forecast, excluding the initial 6 hours of a day-long prediction.

Adding it all together, this is the correct interpretation: Routine forecast for JFK Airport, New York...on the 12th of the month, at 1400Z...valid for 24 hours from 1400Z on the 12th to 1400Z on the 13th...initial temporary condition between 1400Z and 1600Z with visibility at 4 SM due to haze...from 1600Z, winds from 170° at 10 knots, gusting up to 20 knots with visibility beyond 6 SM...a scattering of clouds at 3,500 feet and broken cloud layer at 20,000 feet...leading to a period with winds from 150° at 10 knots, visibility beyond 6 SM, and cloud conditions changing...and so forth until the report's conclusion.

Exercises:

Now, to put your skills to the test, here are three TAFs reports for you to interpret on your own. Use the breakdown structure above to help guide your interpretation. Once you have completed your interpretations, you can review your responses in this guide's "ANSWERS" section.

1. TAF KLAX 221400Z 2214/2314 27010KT P6SM SCT020 BKN050 FM222000 25015KT P6SM SCT025 SCT200 FM230300 28008KT 6SM BR SCT015 BKN030 PROB30 2306/2310 4SM BR BKN015

2. TAF EGLL 221400Z 2214/2314 06008KT 9999 SCT030 BKN080 TEMPO 2218/2222 4000 -RA BKN025 PROB40 2300/2304 3SM RA BKN015 FM230500 08012KT P6SM BKN020 OVC040

3. TAF YSSY 221400Z 2214/2314 04008KT 9999 FEW030 SCT060 FM221800 02010KT 8000 -SHRA SCT020 BKN040 TEMPO 2220/2224 5000 TSRA BKN025CB FM230400 05006KT P6SM SCT030 SCT060

Convective Significant Meteorological Information

Convective Significant Meteorological Information are shared when severe thunderstorms manifest with ground winds surpassing 50 knots, surface hail measuring ¾ inch or more in diameter, or the emergence of tornadoes. Moreover, they are broadcasted to notify pilots about concealed thunderstorms, contiguous thunderstorm chains, or thunderstorms exuding intense or higher precipitation covering at least 40% of an area spanning 3,000 square miles or more. Such meteorological warnings are instrumental for remote pilots during their flight strategizing.

Here is a practical example:
WST 21C WT 232040
AIRMET TANGO UPDT 3 FOR TURB VALID UNTIL 240300
AIRMET TURB...NM TX
FROM 30ESE TBE TO INK TO ELP TO 50S TUS TO 50W LBB TO 30ESE TBE
MOD TURB BTN FL240 AND FL390. CONDS CONTG BYD 03Z THRU 09Z.

Let us break it down and understand what each part refers to:

- WST: Identifies the report as a Convective SIGMET.
- 21C WT 232040: This indicates the issuance number, region (C=Central), and the issue time (23rd day at 2040Z).
- AIRMET TANGO UPDT 3 FOR TURB: This is the AIRMET identifier, signifying it is an update (the third one) about turbulence ("TANGO" is a code for turbulence).
- VALID UNTIL 240300: This indicates the expiration time of the AIRMET, i.e., it is valid until the 24th day at 0300Z.
- AIRMET TURB...NM TX: Specifies that the turbulence is relevant to areas in New Mexico (NM) and Texas (TX).
- FROM 30ESE TBE TO INK TO ELP TO 50S TUS TO 50W LBB TO 30ESE TBE: This defines the boundary of the area affected by turbulence. These are waypoints or navigation aids that define the area's perimeter.
- MOD TURB BTN FL240 AND FL390: Describes the intensity of turbulence (MODerate) and the altitude range (between Flight Level 24,000 feet and Flight Level 39,000 feet).
- CONDS CONTG BYD 03Z THRU 09Z: This suggests that these turbulent conditions are anticipated to continue beyond 0300Z and will persist until 0900Z.

Therefore, the correct interpretation is the following: moderate turbulence between flight levels 24,000 feet and 39,000 feet across a specified area encompassing parts of New Mexico and Texas. The turbulence conditions are expected to continue beyond 03Z through 09Z.

Exercises:
Here are three Convective Significant Meteorological Information reports for you to practice. To determine the right interpretation, turn to the "ANSWERS" section of the book.

1. WST 17E WT 231850

AIRMET SIERRA UPDT 2 FOR IFR AND MTN OBSCN VALID UNTIL 240200
AIRMET IFR...CA NV
FROM 40NNE MOD TO 30SSE BTY TO 40W LAS TO 50N RZS TO 40NNE MOD
CIG BLW 010/VIS BLW 3SM PCPN/BR. CONDS ENDG 00-03Z.

2. WST 09W WT 231720
   AIRMET ZULU UPDT 1 FOR ICE AND FRZLVL VALID UNTIL 240000
   AIRMET ICE...OR WA
   FROM 20W HUH TO 40SSW PDT TO 50SE DSD TO 30N OLM TO 20W HUH
   MOD ICE BTN FL180 AND FL260. CONDS CONTG BYD 00Z THRU 06Z.

3. WST 22S WT 232115
   AIRMET SIERRA UPDT 4 FOR IFR AND MTN OBSCN VALID UNTIL 240400
   AIRMET IFR...CO UT AZ
   FROM 30NNW DEN TO 20SE MTJ TO 40ESE PHX TO 50SSW BCE TO 30NNW DEN
   CIG BLW 005/VIS BLW 2SM BR/FG. CONDS ENDG 02-05Z.

Chapter 3 - Effects of Weather

In this chapter, we delve into the key aspects influencing aircraft performance. These encompass the vehicle's mass, prevailing atmospheric elements, the conditions of the runway, and the fundamental scientific principles that dictate the forces applied to an aircraft.
The effectiveness of an aircraft is significantly influenced by the weather. Hence, a deeper understanding of two primary atmospheric variables—pressure and temperature—is essential.

A Deeper Look at Density Altitude

When understanding aerodynamic performance in non-standard atmospheric conditions, a more appropriate term is "density altitude." This refers to the altitude within a standard atmosphere where a specific air density value is found.

When the air becomes denser (leading to a reduced density altitude), the performance of an aircraft gets a boost. On the other side, the aircraft's performance takes a hit as the air's density decreases (resulting in a higher density altitude).

Factors influencing air density include variations in altitude, ambient temperature, and humidity levels. Dense air is associated with a low-density altitude, while thin air corresponds to a high-density altitude. Situations that might give rise to a high-density altitude include being at lofty elevations, experiencing elevated temperatures, low atmospheric pressures, heightened humidity, or a blend of these elements. In contrast, dense air or low-density altitude is typified by being at lower elevations, encountering colder temperatures, high atmospheric pressures, or reduced humidity.

The Role of Pressure on Density

Air can be compressed or expanded. When you compress air, it allows a more significant amount to fit within a specific volume. If the pressure on a set volume of air drops, this air then spreads out, occupying a more substantial space. A column of air with reduced pressure has a smaller air mass, meaning its density decreases. It's crucial to remember that pressure and density are directly proportionate. Doubling the pressure doubles the density, and if pressure diminishes, so does the density – this is true only when the temperature remains unchanged.

Temperature's Influence on Density

When a substance undergoes a temperature rise, its density typically drops. In contrast, a temperature drop leads to an increase in density. This means air density has an inverse relationship with temperature, which is valid only under constant pressure.

In our atmosphere, both temperature and pressure tend to drop with altitude. Even though these two factors might seem to counteract each other concerning density, the swift decrease in pressure with a rising altitude takes precedence. Thus, aviators should anticipate a decline in density as they ascend.

Humidity's Impact on Density

The sections above have assumed the air to be perfectly dry. In actual scenarios, air always contains some water vapor. While in some situations, the water vapor present might be insignificant, the humidity can play a crucial role in an aircraft's performance under specific conditions. Moist air becomes less dense than dry air because water vapor is lighter than air. As air's moisture content increases, its density drops, which leads to a rise in density altitude and a corresponding decrease in aircraft performance.

The percentage of water vapor in the atmosphere compared to the maximum quantity it can contain at that temperature is referred to as "relative humidity." The ability of air to hold moisture is temperature-dependent: warm air can retain more water than its colder counterpart. While saturated air has a relative humidity of 100 percent, dry air has a relative humidity of zero. While humidity is not the primary factor for determining density altitude or aircraft performance, its contribution cannot be overlooked.

Here are two examples:

- On a Hot Day: Suppose at a high-altitude airport, on a particularly hot day, an aircraft might struggle with takeoff due to the thin air. This scenario is attributed to the high-density altitude caused by the elevated temperatures, even though the physical altitude has not changed.
- On a Humid Day: Consider a scenario where two days have the same temperature, but one day is significantly more humid than the other. On the more humid day, an aircraft might need a longer runway for takeoff because the increased humidity leads to a higher density altitude, affecting the aircraft's performance.

Understanding Aircraft Performance

Aircraft performance is the aircraft's capabilities and efficiency in executing specific functions. These capabilities primarily encompass takeoff and landing distances, climbing rates, maximum altitude, cargo capacity, travel distance, speed, agility, steadiness, and fuel efficiency.

Dynamics of Climbing Performance

An aircraft's weight, altitude, and setup influence the efficiency with which it climbs. The key to efficient climbing lies in the ability of the plane to generate extra thrust or power.

One of the most important aspects affecting an aircraft's performance is weight. An increase in weight necessitates the plane to maintain a steeper angle of attack to sustain a specific altitude and speed. This more vertical angle intensifies the wing's induced drag and the aircraft's overall drag. To counteract the heightened drag, more thrust is required. However, this means there's lesser thrust set aside for ascending. This is why minimizing weight is a priority in aircraft design, given its profound impact on performance parameters. For instance, consider two identical aircraft, but one is carrying a heavier payload. The heavier aircraft would need a longer runway for takeoff, and its climb rate post-takeoff would be slower than its lighter counterpart due to the increased weight affecting its angle of attack and subsequently, its drag.

Additionally, as an aircraft rises in altitude, its required power escalates while the available power diminishes. As a result, the aircraft's ability to climb is reduced the higher it ascends. Picture an aircraft trying to ascend in a mountainous region versus one over the plains. In the former scenario, the thin air at high altitudes means the engines produce less power, and the wings generate less lift. Consequently, the aircraft in the mountainous region might find its climb rate reduced as it gains altitude compared to the aircraft over the plains.

Gauging Atmospheric Pressure

The International Standard Atmosphere offers a reference point for comparing and comprehending aircraft performance data. The benchmark for sea level pressure within this standard is set at 29.92 "Hg, combined with a default temperature of 59 °F. Another unit used to report atmospheric pressure is millibars (mb). To give a perspective, 1 "Hg is roughly equivalent to 34 mb. Therefore, the standard sea level pressure translates to 1,013.2 mb. Common mb pressure values usually fluctuate between 950.0 and 1,040.0 mb. Various meteorological representations, such as surface analysis charts and cyclone data, employ the mb unit.

Given the global distribution of weather monitoring stations, it is essential to normalize barometric pressure readings. This ensures consistency in records and weather reports. Each station modifies its barometric reading to standardize, typically adding about 1 "Hg for every 1,000 feet elevation. For instance, should a station positioned 5,000 feet above sea level record pressure of 24.92 "Hg, its adjusted sea level pressure would be reported as 29.92 "Hg.

Understanding and monitoring these patterns over vast regions helps meteorologists make precise predictions about the movement of weather systems. As a case in point, if a single weather station registers a consistent rise in pressure, it often signifies the onset of clear skies and pleasant weather. On the other hand, a decline or a sharp pressure drop could be a harbinger of deteriorating weather conditions, even intense storms.

Ground Obstructions

Ground-based obstructions can influence wind behavior, often becoming a hidden danger for aviators. Both natural landscapes and artificial structures can disrupt the steady flow of the wind, producing gusts that vary quickly in speed and direction. Examples of these obstructions include skyscrapers in urban areas or tall trees in suburban regions.

The severity of turbulence stemming from these obstructions correlates with the magnitude of the block and the wind's initial speed. Take, for instance, a scenario where a pilot is navigating near a coastal city with tall buildings. While attempting to land, the pilot might face unexpected wind gusts from the wind's interaction with the skyscrapers. These gusts can make the landing more challenging due to the unpredictable wind direction and speed shifts.

Similarly, flying over mountainous terrains can amplify these effects. As wind gracefully ascends the windward face of a mountain, it often aids aircraft in gaining altitude, offering a lifting sensation. However, the leeward or downwind side of the hill presents a contrasting behavior. Here, the wind flows with the mountain's contours, becoming increasingly turbulent. For instance, while flying over the Rockies, a pilot might experience smooth conditions on one side but could encounter rough and potentially dangerous turbulence on the other, especially during fierce winds. This turbulence could push the plane dangerously close to the mountainous terrain, emphasizing the need for pilots to be cautious and informed when navigating such regions.

Low-Level Wind Shear

Wind shear refers to the variations in wind direction or speed across a short distance. For aircraft, encountering such swift changes can lead to intense updrafts, downdrafts, and unexpected shifts in the plane's horizontal trajectory. The risk is magnified at low altitudes, making low-level wind shear especially dangerous because the aircraft is near the ground. Common scenarios inducing low-level wind shear include advancing weather fronts, thunderstorm activity, temperature inversions, and forceful upper-level winds exceeding 25 knots.

The unpredictable nature of wind shear can dramatically alter an aircraft's performance. Imagine an instance where a plane flying against a headwind suddenly faces a tailwind; this sudden shift will decrease its airspeed and performance. On the other hand, transitioning from a tailwind to a headwind boosts airspeed and performance. Pilots must remain vigilant and poised to adapt swiftly to such rapid alterations to ensure the aircraft's stability.

Among the most treacherous forms of low-level wind shear is the microburst. Often linked with rain that evaporates before reaching the ground, this short-lived but powerful downdraft can range 1-2 miles and plunge as deep as 1,000 feet. During its brief 5–15-minute existence, a microburst can introduce downdraft speeds soaring to 6,000 feet per minute and dramatically shift headwinds by 30-90 knots. Such conditions deteriorate performance and trigger intense turbulence and severe wind shifts. For instance, suppose a drone unknowingly ventures into a microburst. In that case, it might first face a surge in performance due to a headwind, which swiftly transitions to debilitating downdrafts and, finally, an accelerating tailwind. This rapid sequence could force the aircraft close to the terrain or cause a crash. Similarly, during an approach for landing, the aircraft might be pushed to the ground before reaching its designated landing spot.

Wind shear's elusive nature poses a silent threat to aviation. Even though there might be reports of its presence, it often goes unnoticed. Therefore, pilots must always stay attentive to the potential of wind shear, particularly when navigating regions prone to thunderstorms or transitioning weather fronts.

Assessing Atmospheric Stability

In a stable atmospheric setting, upward movements find it challenging, causing minor disturbances to fizzle out swiftly. On the other hand, in an unstable atmosphere, minimal vertical currents can escalate, producing turbulence and potential convective phenomena. Such instability may result in pronounced turbulence, expansive steep cloud formations, and severe weather events.

Moisture and temperature together shape atmospheric stability and the consequent weather patterns. For instance, cool and dry air exhibits strong stability, curbing vertical motion, which often manifests in favorable and clear weather conditions. In contrast, the pinnacle of atmospheric instability is reached when the air is warm and humid. This condition is prevalent in tropical zones during summertime. Here, thunderstorms are a daily occurrence due to the inherently unstable nature of the atmosphere.

Understanding Inversions

Usually, as air elevates and spreads in the atmosphere, its temperature drops. But there is an atmospheric exception to this norm. An inversion occurs when, contrary to the typical pattern, air temperature rises with increasing altitude. Such inversion strata are usually thin, stable air layers near the ground. The temperature climbs to a specific level, marking the inversion's peak. This top layer acts like a barrier, entrapping weather elements and pollutants beneath it. When the air has high relative humidity, it can spur the development of clouds, fog, haze, or even smoke, hampering visibility within this inversion layer.

Inversions near the surface manifest on clear, chilly nights. The ground's dropping temperatures cool the adjacent air, causing the air just above the surface to become cooler than the layers above. Meanwhile, frontal inversions emerge when a layer of warm air overlays cooler air or when colder air is thrust beneath warmer layers.

Interplay of Temperature and Dew Point

The dew point, denoted in degrees, represents the temperature at which air has reached its water vapor capacity. When air's temperature descends to the dew point, it becomes fully saturated. Consequently, water vapor settles out, materializing as fog, dew, frost, clouds, rain, or even snow.

Ways Air Achieves Saturation

When the air attains its saturation level, especially if the temperature and dew point are identical, phenomena like fog, low-hanging clouds, or precipitation are probable outcomes. There exist four primary pathways for air to hit saturation:

- As warm air traverses a cooler surface, it cools down and might hit its saturation point.
- The intermingling of cold and warm air can lead to saturation.
- On nighttime occasions, when air cools by interacting with the chillier ground beneath, it might achieve saturation.
- When atmospheric dynamics push air upwards, it could reach saturation.

Dew and Its Frosty Counterpart

During serene, clear, and cold nights, the temperature of the earth and surface objects can cause nearby air temperatures to fall beneath the dew point. As a result, water vapor condenses and settles on terrestrial features such as structures, cars, aircraft, and even the earth itself. This settled water vapor is called "dew," often visible on vegetation and surfaces during early morning hours. However, when temperatures drop below freezing, this moisture crystallizes as frost. While dew does not impact a small Unmanned Aircraft, ice presents genuine aviation risks. Frost hampers the aerodynamics of the wing, reducing lift generation and augmenting drag. This combination of diminished lift and increased drag can negatively impact the takeoff capabilities of an Unmanned Aircraft. It is imperative to ensure Unmanned Aircrafts are meticulously cleaned and devoid of any frost before initiating any flight.

Cloud Formations

The cumulonimbus cloud stands out as particularly dangerous for pilots. These clouds can be solitary or in clusters, categorized into air mass and orographic thunderstorms. The former emerges due to the Earth's surface heating the adjacent air, while the latter results from the upward drift of air in mountain areas. When cumulonimbus clouds align continuously, they form non-frontal thunderstorm bands, often called squall lines.

These clouds are a product of ascending air currents, so their turbulent nature makes them a significant flight risk. To better understand the concept, a small Unmanned Aircraft navigating a thunderstorm might be subjected to vertical air movements exceeding 3,000 feet per minute.

Thunderstorms can also unleash large hail, powerful lightning, tornadoes, and copious amounts of rain, all of which can jeopardize aircraft safety.

Lenticular Altocumulus Clouds in a Standstill

These clouds emerge atop wave peaks generated by obstructions in wind pathways. They remain stationary—hence, "standing" lenticular. While the clouds might seem stagnant, the winds coursing through can be notably vigorous. Distinguished by their sleek, lustrous edges, their appearance is a warning sign for pilots of potential intense turbulence, urging them to steer clear.

Atmospheric Stability and Its Traits

The inherent stability of an air mass defines its general weather patterns. As one air mass overlies another, characteristics evolve vertically. Here is a comparison of traits associated with unstable and stable air masses:

- Clouds with a cumuliform structure vs Layered clouds and fog
- Precipitation that's intermittent and intense vs Steady rainfall
- Turbulent atmospheric conditions vs Calm atmospheric conditions
- Clarity in visibility (bar exceptions like dust storms) vs Moderate to poor visibility due to factors like haze or smoke.

Navigating Mountainous Regions

Before embarking on a journey across hilly landscapes, procuring in-depth preflight data encompassing cloud observations, air stability, wind patterns, and speed is crucial. Modern tools like satellites can be instrumental in identifying mountain waves. However, sometimes comprehensive data might be elusive, underscoring the need for vigilance during flight. As a rule of thumb, wind speeds surpassing 25 knots at mountain peak levels signify potential turbulence, while those exceeding 40 knots call for heightened caution. Layered cloud formations hint at stable air conditions. The appearance of standing lenticular or rotor clouds is indicative of mountain waves. Thus, anticipate turbulence when flying leeward of mountains but smoother conditions windward. On the other hand, the emergence of convective clouds on a mountain's windward side implies unstable air, signaling turbulence both adjacent to and on either side of the hill.

Structural Icing

For structural icing to occur mid-flight, two prerequisites must be met:

- The aircraft should traverse through visible water sources, like rain or cloud particles.
- The temperature at the contact point, where moisture interacts with the aircraft, should be 32° F or below.

Notably, due to aerodynamic cooling, an aircraft's airfoil can plummet to 32° F, even if the surrounding ambient temperature is slightly warmer.

The Lifecycle of Thunderstorms

The evolution of a thunderstorm can be divided into three distinct phases: (1) the initial growth or cumulus phase, (2) the peak or mature phase, and (3) the concluding or dissipating phase. While observing from the ground, it is challenging to pinpoint the exact transition between these stages, given the smooth progression.

1. The Cumulus Phase

Thunderstorms originate in cumulus clouds, even though not all cumulus clouds become storms. An updraft, or upward-moving air current, is the defining feature of this phase. This updraft can be strong, sometimes elevating clouds at rates surpassing 3,000 feet per minute. This makes it dangerous for small Unmanned Aircraft to operate in regions with rapidly ascending cumulus clouds. A real-life example might be a drone operator who notices a rapidly growing cumulus cloud and wisely decides to ground his equipment. As the cloud develops, tiny water droplets merge to form raindrops. These raindrops are uplifted above the freezing point, presenting potential icing risks. As these raindrops get heavier, they descend, pulling cold air with them, signifying the onset of the mature phase.

2. The Mature Phase

When precipitation starts to emanate from the cloud base, it indicates the commencement of the mature phase. The falling cold rain limits the warming effect of compression on the downdraft, making it cooler than the surrounding air. As a result, the downward speed intensifies, surpassing 2,500 feet per minute. This descending air disperses out when it reaches the surface, leading to powerful gusty winds, a noticeable drop in temperature, and a swift surge in atmospheric pressure. This gust front can be likened to a sudden cold blast at a picnic, causing paper plates and napkins to fly off. The mature phase is also marked by the strongest updrafts, reaching velocities greater than 6,000 feet per minute. The proximity of these updrafts and downdrafts generates a turbulent environment, with the mature phase being the most hazardous for flight operations.

3. The Dissipating Phase

This final stage is characterized by downdrafts, signaling the decline of the storm. Once the rainfall ceases and the downdrafts diminish, the storm can be considered to have entered the dissipating phase. For instance, after a short-lived but intense storm during a baseball game, players and spectators might notice the sky clearing up and conditions becoming more tranquil. Only benign cloud remnants linger after all the storm's cells have transitioned through this phase.

Cloud Cover and Altitude

In aviation terminology, a 'ceiling' denotes the lowest cloud layer that is broken or overcast, or it can refer to the visual vertical range through obstructions such as fog or haze. A sky is termed 'broken' when it is 5/8 to 7/8 covered by clouds. When clouds entirely mask the sky, it is labeled 'overcast.' Aviators turn to the METAR and other automated weather monitoring systems to get the latest on ceiling conditions.

Visibility

Hand in hand, cloud cover, and ceiling data is the crucial visibility metric. The maximum horizontal distance at which recognizable landmarks or objects can be seen by the unassisted eye is known as visibility. METARs and other aviation-specific weather reports and automated weather platforms provide current visibility readings. Additionally, pilots can obtain forecasted visibility conditions during their preflight weather briefings.

Chapter 4 - Mass, Stability, Load Factors and Weight Control

For every flight operation, the drone operator or pilot-in-command should ensure the aircraft's weight and balance by setting guidelines by inspecting its load conditions. Adherence to the weight and balance specifications, as defined by the manufacturer or the builder, is crucial for a safe flight experience. The operator should recognize the risks of flying an overweight aircraft, especially if unforeseen emergencies arise.

There may be a defined maximum gross takeoff weight, but that does not always guarantee a safe lift-off under all circumstances. Factors such as high terrain elevations, increased air temperatures, and elevated humidity levels (resulting in high-density altitudes) might require weight adjustment before attempting to fly. The pilot must also evaluate the launch area's length, condition, inclination, prevailing wind conditions, and any potential obstructions. All these elements could necessitate weight adjustments before initiating flight.

The performance of an airplane is significantly influenced by weight changes while in flight. Fuel use is the most typical illustration of this. As the aircraft utilizes fuel, it becomes lighter, potentially enhancing its performance. However, this can conversely impact the aircraft's balance. In drone operations, other onboard items, like a releasable load, can also change the aircraft's weight during a flight.

Balance issues related to weight distribution can influence flight characteristics similarly to those arising from excessive weight. The manufacturer often determines boundaries for the center of gravity (CG). The center of gravity is not a stationary point but varies depending on how weight is distributed across the aircraft. Changing or using variable load items can shift the center of gravity, and the pilot-in-command needs to anticipate these shifts and their potential effects on the aircraft. If the center of gravity is not within safe limits post-loading or moves out of these limits during the flight, it might need to adjust or offload some weight before initiating the flight.

Mass and Balance

Everything is drawn toward the center of the Earth by gravity. Think of the Center of Gravity as the singular point where an aircraft's entire weight is focused. If one were to suspend the aircraft from its precise center of gravity, it would remain balanced regardless of its orientation. The center of gravity plays a pivotal role in drones and other small crewless aircraft, significantly impacting stability. The aircraft's unique design dictates the center of gravity's permissible position. The size of the center of pressure (CP) movement is measured by aircraft designers. It is vital to comprehend that while an aircraft's weight converges at the center of gravity, the lift's aerodynamic forces manifest at the center of pressure. When the center of gravity is higher than the center of pressure, an aircraft would often tilt downward. In contrast, if the center of pressure is ahead of the center of gravity, a nose-up tendency develops. Thus, to maintain flight balance, designers establish the center of gravity's rear limit ahead of the center of pressure for the corresponding flight velocity.

Weight and lift clearly have a relationship. The upward force acting on the wing is known as lift, and it is perpendicular to both the incoming wind and the side axis of the aircraft. It counteracts the aircraft's weight. In steady, level flight, the aircraft achieves equilibrium when the lift matches its weight, preventing upward or downward acceleration. If lift diminishes below the weight, the vertical speed will drop. On the other hand, if lift surpasses weight, the vertical ascent rate will amplify.

Understanding Aircraft Stability

Stability refers to an aircraft's ability to rectify any disruptions to its balance and realign with its initial trajectory. This quality is fundamentally rooted in the design of the aircraft.

Two primary aspects influenced by an aircraft's stability are maneuverability and controllability. Maneuverability defines how effortlessly an aircraft can be directed and its capacity to endure the strains of these movements. Factors such as the aircraft's weight, inertia, design and placement of flight controls, structural durability, and its power source play crucial roles. This too emerges from the aircraft's design.

Controllability pertains to an aircraft's responsiveness to pilot directives, particularly concerning its flight direction and posture. It reveals how the aircraft reacts to piloting inputs, independent of its inherent stability.

Load Factors

The peak load factor, relevant to a specific bank angle, is a ratio of lift to weight interlinked through trigonometry. The load factor is quantified in gravitational acceleration units (Gs), representing the force comparable to the gravitational pull on a stationary object. This reflects the intensity of force an object undergoes when accelerated. Every force propelling an aircraft from a linear path imposes a structural stress termed the load factor. While delving deep into aerodynamics is not obligatory for acquiring a remote pilot license, pilots must grasp the forces at play, their strategic utilization, and the operational constraints of their aircraft.

To illustrate the concept, a load factor of 3 implies the overall stress on an aircraft's framework is triple its weight. As these factors are articulated in Gs, a three-load factor can be denoted as 3 Gs.

Grasping load factors is pivotal for two core reasons:

1. Pilots possess the potential to subject the aircraft structures to hazardous overloads.
2. Elevating the load factor augments the stalling speed, thereby rendering stalls plausible even at speeds perceived as safe.

Here is an illustration to help you comprehend this idea. Think of a play drone. If you make it turn or maneuver too sharply, you can observe how it strains to maintain its balance. This is like load factors in actual aircraft. If you had a drone designed to handle up to 3 Gs and tried to make it maneuver under 4 Gs, you risk damaging it or causing it to stall even if it seems to be flying at a safe speed.

Understanding Aircraft Weight and Balance

Respecting the aircraft's weight and balance thresholds is paramount for the safety of any flight. Surpassing the stipulated maximum weight endangers the aircraft's structural soundness and negatively impacts its performance. Flying with the center of gravity outside the designated limits can lead to challenging control situations.

Weight Control

Weight is the gravitational pull exerted on an object, drawing it towards Earth's core. This force emerges from the interplay between an object's mass and the gravitational acceleration it experiences. Weight plays a vital role in an aircraft's design and operation and necessitates due diligence from pilots. The persistent tug of gravity aims to drag the aircraft downwards. The opposing force, lift, negates this pull and maintains the aircraft in flight. However, an airfoil's lift generation is bound by its design, angle of attack, speed, and atmospheric density. It is crucial to ensure that the aircraft is not loaded beyond the manufacturer's guidelines, for if the weight surpasses the lift, the aircraft might be rendered flightless.

Impact of Excess Weight

Additional weight added to an aircraft, detracting from its overall weight limit, could be better for its performance. Aircraft manufacturers prioritize crafting lightweight designs without compromising on durability or safety standards. Pilots should remain vigilant about the repercussions of overburdening their aircraft. An overloaded craft might struggle to ascend, and even if it manages to do it, it could display unpredictable and compromised flight behaviors. The first sign of a crowded aircraft often emerges during the takeoff phase.

Overburdening an aircraft result in notable declines in its performance metrics. Some of the critical performance drawbacks of an overweight aircraft include:

- Need for increased takeoff velocity

- Extended takeoff distance
- Diminished climb rate and trajectory
- Capped maximum altitude
- Limited flight range
- Slowed cruising pace
- Hampered agility
- Augmented stalling velocity
- Elevated landing and approach speeds
- Prolonged landing duration

Think of an aircraft like a teeter-totter. On one end, you weight the aircraft, pulling it down; on the other, the lift is trying to raise it. Adding too much weight to one side (like overloading the teeter-totter with too many children) makes it harder for the other side (or the lift) to balance it out. This makes taking off, cruising, and landing more difficult and dangerous.

Beyond its inherent risks, excess weight further narrows the safety buffer, especially when combined with other performance-limiting elements. In emergencies, the implications of an overloaded aircraft can be dire.

Chapter 5 - Inflight Emergencies

Emergencies can catch a remote pilot off guard, potentially leading to severe outcomes. Should an emergency arise during a flight, remote pilots can deviate from the regulations laid out in 14 CFR part 107 to address the situation effectively. If the FAA requests, a remote pilot who has deviated from the rules due to an emergency must report the incident.

Navigating Inflight Emergencies

The onus of ensuring the small, unmanned aircraft's safety during operations lies with the remote pilot. This includes guaranteeing the aircraft's soundness for flight, confirming that it does not pose risks to individuals or property, and ensuring that all participating crew members are well-informed about standard and emergency procedures.

A thorough aircraft preflight inspection is mandatory for a remote pilot before each flight. Any anomalies or issues detected during this check must be addressed before initiating the flight with the small, unmanned aircraft. While some manufacturers equip remote pilots with a detailed preflight checklist, in its absence, it is prudent for the pilot to create a comprehensive list, ensuring the aircraft's readiness for safe operation.

If an inflight emergency arises, the remote pilot's priority is to minimize risks to individuals and property. For instance, should the small, unmanned aircraft encounter a battery fire during flight, the pilot might need to elevate its altitude above 400' AGL to navigate to an appropriate landing zone safely. In such scenarios, the pilot is required to submit a report if solicited by the FAA.

It is essential that when auxiliary crew members are involved in a flight operation, they are comprehensively briefed both on the flight details and the designated emergency protocols. This briefing should be extended to any visual observers participating and to any individual who, despite not being certified, is granted permission to operate the small, unmanned aircraft's flight controls.

Effective Scanning Methods

Pilots should transition their gaze from right to left or vice versa for efficient scanning. Start the scan from the farthest point where an object can be discerned (at the outermost range) and progress towards the aircraft's current position (closer range). The scan should encompass a visual field about 30° in width at each interval. While the time taken for each interval can vary based on the intricacy of details needed, it is recommended not to exceed 2 to 3 seconds for any interval. As pilots shift from one visual interval to the next, the previously viewed area should overlap by approximately 10°.

Chapter 6 - Radio Communication Practices

Secure aircraft operations within the National Airspace System (NAS) rely on proficient radio communications. This channel facilitates the relay of vital details before, amidst, and after a flight and plays an instrumental role in streamlining aircraft movement in dense and sparse airspace regions. Radio communications also serve as the backbone for addressing critical flight safety issues, encompassing unexpected meteorological challenges and inflight adversities. While radio frequency communications might not be the norm for small, unmanned aircraft pilots, comprehending aviation-specific terminologies and the myriad of dialogues they might stumble upon is paramount, mainly if they rely on radios for heightened situational awareness within the National Airspace System. While the content herein predominantly caters to pilots of human-crewed aircraft, unmanned aircraft pilots must grasp the distinct nuances of information dissemination in the National Airspace System.

Airport Activities in Absence of Control Towers

While navigating near an airport, maintaining vigilance is paramount. This becomes especially vital when operating at an airport lacking a control tower, given that some aircraft might need more communication equipment. Others might have not announced their intentions or presence when approaching or departing these airports. Ensuring safety demands that all aircraft equipped with radios communicate on a shared frequency, with small, unmanned aircraft pilots tracking communications for situational awareness.

Airports may possess a part-time or full-time tower or an on-site flight service station (FSS). As an alternative, there might be no aeronautical station, a full-time or part-time Universal Communications (UNICOM) station, or both. Pilots have three communication options at such airports: interacting with an FSS, a UNICOM operator, or making self-announce broadcasts. Today, many airports use automated UNICOM systems that deliver weather updates, radio checks, and advisory information. Both the approach charts and the airport/facility directory allow you to check the status of these systems.

Suggested Traffic Advisory Practices

Although not mandated for remote pilots to communicate with manned aircraft near non-towered airports, understanding traffic patterns and radio procedures enhances safety. Before operating near such an airport, identify relevant frequencies. Most have a typical UNICOM frequency, like 122.8. If no UNICOM or other frequencies are available, use MULTICOM 122.9.

As manned aircraft approaches a non-towered airport, they typically announce their presence when 10 miles out, indicating their relative position (north, south, east, or west).

Example: "Hillview traffic, Piper 456 Delta Charlie is 10 miles north, approaching for landing, Hillview traffic."

Subsequent broadcasts should be like: "Hillview traffic, Piper 456 Delta Charlie, entering pattern, mid-field right down-wind for runway 24, Hillview traffic."

For aircraft making direct landings, especially during instrument approaches, the message might be: "Hillview traffic, Piper 456 Delta Charlie, two miles west, direct approach runway 24, full stop, Hillview traffic."

As they navigate the landing pattern, pilots should continuously update their positions:

- "Hillview traffic, Piper 456 Delta Charlie, right base, runway 24, Hillview traffic."
- "Hillview traffic, Piper 456 Delta Charlie, on final, runway 24, Hillview traffic."

After landing and exiting the runway: "Hillview traffic, Piper 456 Delta Charlie, has cleared runway 24, heading to hangars, Hillview traffic."

For departing aircraft, a typical broadcast might be: "Hillview traffic, Piper 456 Delta Charlie, taking off from runway 24, Hillview traffic."

Remote pilots should always monitor their surroundings, especially near airports. While manned aircraft ideally announce their intentions, it is not a regulatory requirement. Thus, for added situational awareness, using a radio is beneficial.

Recognizing Aircraft Call Signs

Aircraft call signs are unique identifiers, essential for clear communication between pilots and air traffic controllers (ATCs). In the realm of commercial airlines, call signs typically merge the airline's name or its abbreviation with the flight number. For example, a Delta Air Lines flight 123 would use the call sign "Delta 123".

For general aviation, which includes private aircraft, the aircraft's registration number is typically its call sign. In the U.S., these registrations often start with an 'N' followed by a combination of numbers and letters. An aircraft with the registration N123AB would communicate using the call sign "November One Two Three Alpha Bravo", pronouncing each digit and letter distinctly for clarity.

Military aircraft have their own system, often employing specific call signs based on their mission, unit, or aircraft type. These can be very varied and might not relate to the aircraft's registration or model.

There are also exceptional cases to consider. Some aircraft, engaged in unique operations or missions, might adopt specific call signs relevant to their function. For instance, when the U.S. President is aboard, the aircraft is commonly known by its call sign "Air Force One", even though its technical name is based on its registration.

Accuracy in pronunciation is pivotal. Numbers should be articulated clearly, like using "niner" for the number 9 to prevent confusion with similar sounding words. To promote clarity in communication, letters are also pronounced utilizing the phonetic alphabet.

In certain scenarios, especially during extended flights, aircraft may change their call signs. This often happens when transitioning between different airspaces or under different controlling entities, and pilots will inform the relevant ATC of any change.

The primary objective behind using call signs is to maintain safety. In busy airspace, having a distinct identifier for each aircraft is crucial. Moreover, concise call signs improve communication efficiency and reduce the chance of errors.

Chapter 7 - Physiological Factors and Pilot Performance

According to 14 CFR part 107, drone operations are prohibited if the remote pilot-in-command, the individual handling the drone's controls, or the Visual Observer is not able to execute their duties safely. The onus is on the remote pilot-in-command to ensure that no operational team member is under any kind of impairment.

While the detrimental effects of drugs and alcohol on judgment are well-recognized, even some over-the-counter medications could compromise the safe handling of a drone. For instance, certain types of cold medicines and antihistamines can induce sleepiness. Moreover, it is crucial to note that part 107 explicitly forbids individuals from undertaking roles if:

- They have consumed alcohol in the last 8 hours.
- They are presently influenced by alcohol.
- Their blood alcohol concentration is at or exceeds .04 percent.
- They are on medication that might impair their mental or physical capacity.

Furthermore, specific health conditions, like epilepsy, can pose operational risks. It is up to the remote pilot-in-command to assess their health condition, ensuring it is managed appropriately and does not hinder their ability to operate a drone safely.

Health and Physiological Factors Influencing Pilot Capabilities

Pilots need to be aware of specific health concerns that can influence their performance. Key factors to consider are:

- Over-breathing
- Tension
- Exhaustion
- Lack of hydration
- Overheating
- Influence of substances and alcohol

Over-breathing

Over-breathing, medically termed hyperventilation, is the intensified pace and depth of breathing, which results in an atypical decrease of carbon dioxide in one's bloodstream. This situation is more prevalent among pilots than many assume. While it rarely leads to total incapacitation, it can generate symptoms that can concern pilots who are not familiar with it. This unfamiliarity, combined with heightened breathing and anxiety, can exacerbate the situation. An unforeseen stressful event can unconsciously prompt pilots to amplify their breathing pace.

Frequent indicators of over-breathing encompass:

- Vision issues
- Potential fainting
- A sensation of light-headedness or dizziness
- Numb or prickly feelings
- Sensations of varying temperature
- Muscle contractions

Managing over-breathing revolves around reestablishing the body's carbon dioxide balance. Breathing regularly is the most effective prevention and remedy for this condition. To counteract over-breathing, one can reduce breathing speed, breathe into a paper bag, or speak aloud. Once breathing stabilizes, recovery is typically swift.

Stress

Stress can be described as the body's reaction to various mental and physical demands. When confronted with stress, the body secretes specific hormones, like adrenaline, into the bloodstream and accelerates its metabolism, fueling the muscles with more energy. As a result, there is a surge in blood sugar, heart rate, breathing rate, blood pressure, and sweat production. A "stressor" is a factor that induces stress in an individual. Stressors can range from physical elements (like loud noises or vibrations) to physiological ones (such as tiredness) and even psychological triggers (challenging tasks or personal challenges).

Stress is classified into two main types: acute and chronic. Acute stress arises from immediate perceived threats, evoking the "fight or flight" reflex, regardless of whether the threat is tangible. Under normal circumstances, a healthy individual can manage acute stress without becoming overwhelmed. Yet, if acute stress persists, it can transition into chronic stress.

Chronic stress represents a degree of overwhelming stress that goes beyond an individual's coping capacity, causing a significant drop in performance. Continuous emotional strains, be it from solitude, financial concerns, or issues related to relationships or work, can cumulatively generate a stress level that becomes too much for someone to handle. At such elevated stress levels, a pilot's performance can deteriorate drastically. Such pilots are unfit to fly and should refrain from using their flying rights. If a pilot believes they are grappling with chronic stress, seeking medical advice is essential.

Fatigue

Fatigue is a common cause of pilot errors. Symptoms of fatigue encompass reduced attention and focus, diminished coordination, and hampered communication ability. All these can compromise decision-making abilities. Physical fatigue emerges from sleep deprivation, vigorous exercise, or intense physical labor. On the other hand, mental fatigue often stems from stress and extended periods of cognitive tasks.

Acute and chronic fatigue can be broadly divided into two categories. Acute fatigue is temporary and is a regular part of daily life. It is exhaustion after intense activities, heightened emotions, or insufficient sleep. Typically, relaxation post-activity and a solid 8-hour sleep can alleviate this fatigue.

A specific form of acute fatigue is skill-related fatigue. This manifests in two primary ways:

- Timing disruption - where one appears to carry out a task normally, but the synchronization of each component is slightly misaligned. This leads to a segmented, rather than a seamless execution of tasks.
- Perceptual field disruption - which is characterized by a focus on central visual tasks and neglect of peripheral ones, resulting in less precise control movements.

Several factors can induce acute fatigue in pilots, including:

- Mild oxygen deprivation
- Physical pressures
- Emotional tensions
- Physical energy drains due to emotional strains
- Continuous emotional burdens

Maintaining a balanced diet and ensuring sufficient rest is crucial to prevent acute fatigue. A wholesome diet prevents the body from using its tissues for energy, and adequate relaxation conserves its energy reserves.

On the other hand, chronic fatigue spans over an extended duration and often has psychological origins, though it can also arise from persistent medical conditions. Persistently elevated levels of stress are a leading cause of chronic fatigue. Unlike acute fatigue, chronic fatigue is not typically alleviated by regular diet or rest. Symptoms may manifest as feelings of exhaustion, heart palpitations, shortness of breath, migraines, or mood swings. At times, this can also lead to gastrointestinal issues and body aches. If left unchecked, it can evolve into emotional disorders.

Remote pilots should refrain from operating a small, unmanned aircraft when experiencing acute fatigue. The onset of fatigue during a flight operation cannot be countered by training or expertise. The sole remedy to fatigue is ample rest. Pilots should only operate an unmanned aircraft with proper rest, especially after long work hours or particularly draining or stressful days. Medical advice is essential for those suspecting they might be dealing with chronic fatigue.

Dehydration

Dehydration can result from elevated temperatures, exposure to wind humidity, and consumption of diuretics like coffee, tea, alcoholic beverages, and caffeine sodas. Typical symptoms of dehydration include headaches, tiredness, muscle cramps, drowsiness, and light-headedness.

One of the first evident effects of dehydration is tiredness, which can impact mental and physical performance. Operating a small, unmanned aircraft for extended durations in warm summer climates or elevated altitudes amplifies the risk of dehydration. This is due to the accelerated loss of water under such conditions.

As a preventive measure against dehydration, consuming between two to four quarts of water daily is advisable. However, individual needs might vary due to physiological differences. A common recommendation recommends consuming two quarts, or eight 8-ounce glasses of water, each day. Not replenishing this lost fluid can lead to more severe symptoms like dizziness, weakness, nausea, tingling sensations in extremities, stomach cramps, and intense thirst.

It is crucial for pilots to monitor their hydration levels constantly. When there is a deficit of about 2% in body weight, people experience thirst. This is the body's innate mechanism, signaling the need for hydration. The challenge, however, is that this mechanism is delayed in its activation and can be prematurely deactivated by just a tiny quantity of fluid intake, leading to further delay in adequate rehydration.

Some proactive measures to counter dehydration are:

- Regularly monitoring and measuring fluid intake using a dedicated container.
- Proactively hydrating rather than waiting for thirst signals.
- If one is not inclined towards drinking plain water, consider adding flavors or opting for hydration solutions.
- Minimizing daily consumption of caffeine and alcohol, as they promote fluid loss through increased urination.

Overheating

Overheating is a severe condition where the body cannot regulate its temperature. Initial signs might be akin to dehydration, though, in some instances, it is only identified when someone collapses suddenly.

Even though you do not feel thirsty, it is imperative to consistently hydrate in order to avoid such difficulties. The body typically processes about 1 to 1.5 quarts of water an hour. In intense heat situations, it is advisable to consume about a quart of water every hour, and in less demanding conditions, a pint per hour is sufficient.

Drugs

While the Federal Aviation Regulations do not lay out specifics on drug consumption, Title 14 of the Code of Federal Regulation does state that anyone:

- Aware (or suspects) of a medical condition that may hinder the ability to meet necessary medical standards for piloting should abstain from flying.
- Undergoing treatment or medication that can impact their ability to meet medical standards should also avoid piloting.

14 CFR part 107 as well as sections 91.17 and 91. of 14 CFR part 91.19, emphasize the prohibition of drugs that can compromise one's abilities, endangering flight safety.

There are numerous FDA-approved medications, excluding over-the-counter drugs. Most drugs can exhibit adverse reactions in specific individuals. Herbal remedies, dietary supplements, and some "organic" products can also cause adverse side effects, like standard medications. Hence, the FAA consistently evaluates data to ensure that acceptable drugs for pilots do not pose significant safety concerns.

Many over-the-counter drugs, like antihistamines and decongestants, can produce marked side effects, such as fatigue or cognitive impairments. While colds and respiratory issues might deter pilots from flying, combining them with drugs that induce side effects exacerbates the problem. For instance, medicines like Benadryl cause sleepiness and remain in the system for an extended period, prolonging the duration of potential side effects.

Before every flight, pilots must undergo a thorough self-check to ensure they are fit for the skies. A helpful acronym is IMSAFE: Illness, Medication, Stress, Alcohol, Fatigue, and Emotion. When evaluating the Medication aspect, pilots should question, "Have I consumed any drug that might cloud my judgment or induce sleepiness?" After taking any new medicine, be it over the counter or prescription, there is a recommended 48-hour window before piloting to ascertain the absence of any adverse side effects.

Alongside medication considerations, pilots should also:

- Avoid non-essential drugs.
- Stick to a balanced diet.
- Pack light snacks.
- Ensure proper hydration, with sufficient water onboard.
- Prioritize good sleep the night before flying.
- Maintain physical fitness for optimal performance.

The Impact of Alcohol on Piloting

The success and safety of a flight hinge upon the capacity to make correct judgments and adopt appropriate measures in regular and unexpected circumstances. The introduction of alcohol into this equation significantly diminishes the likelihood of a flight concluding without mishap.

Even small quantities of alcohol can compromise judgment, reduce a sense of duty, hinder coordination, narrow the field of vision, weaken memory, curtail logical thinking, and diminish concentration. A mere ounce of alcohol can slow down and cut muscle reflexes, reduce the effectiveness of eye movements, especially during reading, and elevate the chances of making mistakes. Consuming just one alcoholic beverage can lead to compromised vision and hearing.

Moreover, the effects of alcohol linger. When a person has a hangover, they still grapple with alcohol's effects. Even if a pilot feels they are operating at peak efficiency, the reality is that their cognitive and physical responses are still hampered. Because alcohol can stay in the bloodstream for up to 16 hours, pilots should use caution when flying soon after drinking. Alcohol intoxication is gauged by its concentration in the bloodstream, often represented as a percentage.

In accordance with 14 CFR part 91, pilots must keep their blood alcohol level under .04 percent, and there should be an interval of at least 8 hours between alcohol consumption and flying. If a pilot's blood alcohol concentration is .04 percent or higher, even after waiting for 8 hours, they are prohibited from flying until their blood alcohol concentration drops beneath that threshold. Despite the blood alcohol concentration being below .04 percent, flying within 8 hours of alcohol consumption remains forbidden. Even though these guidelines are clear-cut, it is prudent for pilots to adopt an even more cautious stance than what is mandated.

Chapter 8 - Evaluating, Addressing and Managing Risk

Aeronautical decision-making represents a deliberate methodology pilots employ to consistently evaluate and decide on the most appropriate actions given specific scenarios. Aeronautical decision-making encapsulates a pilot's choices based on the most recent data available.

Emphasizing the significance of mastering robust aeronautical decision-making techniques is vital. While we observe ongoing advancements in pilot training techniques, enhancements in aircraft tools and systems, and increased provisions for pilots, accidents still transpire. Even with the technological strides to bolster flight safety, one element still needs to be made public: human susceptibility to errors. Around 80% of all aviation mishaps are projected to stem from human-related factors, with a considerable portion of these incidents transpiring during landing (about 24.1%) and takeoff (23.4%).

Aeronautical decision-making can be viewed as a structured strategy for gauging risks and managing stress. Recognizing aeronautical decision-making also entails understanding the role of personal biases in shaping decisions and recognizing that these predispositions can be adjusted to ensure the safe operation of small, unmanned aircrafts further. It is pivotal to be aware of the dynamics leading to human decisions and to comprehend the mechanics of the decision-making process and the ways it can be refined.

The Role of Human Elements in Aviation

Why is it pivotal to recognize human states such as exhaustion, overconfidence, and anxiety in aviation? These, among others, are encapsulated under the term 'human elements.' It is alarming that these human elements either directly instigate or contribute significantly to many aviation mishaps. Over 70% of aircraft incidents have had human elements as a principal factor.

Although often linked to flight operations, concerns about human elements have recently expanded, encompassing aviation maintenance and air traffic coordination. Considering this, the FAA has intensified its focus on understanding and researching human factors over recent years. This has involved close collaborations with professionals such as engineers, pilots, mechanics, and air traffic controllers. The goal? To incorporate the latest insights on human elements, facilitating better safety and operational efficiency.

The discipline of human elements is multifaceted, drawing from diverse fields such as psychology, engineering, design, statistics, operations analytics, and human body measurement. It is about comprehending human capabilities and applying this knowledge to conceptualizing, creating, and implementing systems and services. Moreover, it is about integrating these principles across various aviation dimensions, including pilots, air traffic controllers, and maintenance personnel.

While some might equate human elements with Crew Resource Management (CRM) or Maintenance Resource Management, the reality is that the scope of human factors is much broader. It delves into specific research related to varied contexts (e.g., flight conditions, maintenance scenarios, stress thresholds, expertise) about human capacities, restrictions, and other traits. This research subsequently informs the creation of tools, machinery, systems, tasks, roles, and settings that align with human needs and abilities, ensuring safety and ease of use. The aviation sector stands to gain tremendously from advancements in human elements, offering insights that can enhance the way professionals engage with their roles and tools.

The Evolution of Aeronautical Decision-Making

For over a quarter-century, stellar pilot judgment, often termed aeronautical decision-making, has been acknowledged as pivotal for aircraft safety and for averting accidents. Driven by the urge to curtail accidents stemming from human errors, the airline sector pioneered training modules rooted in aeronautical decision-making enhancement. Crew Resource Management training, catered to flight crews, zeroes in on harnessing all accessible resources: human capital, equipment, and pertinent information. This is done to foster team synergy and elevate decision-making capabilities. Every flight crew aspires for adept aeronautical decision-making, and Crew Resource Management serves as a tool to promote sound decisions.

This burgeoning interest in decision-making propelled the Federal Aviation Administration to craft training to refine pilots' decision-making skills. This initiative eventually informed the current FAA mandates, which necessitate the inclusion of decision-making in pilot training modules. The FAA must quiz an applicant on Aeronautical Decision Making and Risk Management for the acquisition of an sUAS certificate. By 1987, this focus on aeronautical decision-making crystallized with the launch of six manuals tailored to cater to the decision-making essentials of pilots with varied ratings. These comprehensive guides were crafted with the aim of curtailing accidents tied to decision-making missteps. Independent evaluations affirmed the efficacy of these resources when novice pilots, underpinned by this training along with the regular flight curriculum, manifested marked improvement. Pilots equipped with aeronautical decision-making knowledge showcased 10 to 50 percent fewer in-flight errors than their counterparts without aeronautical decision-making training. In real-world settings, an operator clocking 400,000 flight hours annually recorded a 54 percent dip in accident rates post-incorporation of these resources in refresher training sessions.

Challenging the prevalent notion, superior judgment is indeed teachable. The age-old belief tethered good judgment as a spontaneous outcome of amassed experience. It was widely accepted that their judgment skills would naturally enhance as pilots accrued accident-free flight durations. Building on this traditional decision-making foundation, aeronautical decision-making fine-tunes this process to diminish the chances of human discrepancies and augments the likelihood of incident-free flights. Aeronautical decision-making offers a methodical, structured lens to scrutinize any shift during a flight and its ramifications on flight safety. The aeronautical decision-making mechanism delves deep into the decision-making spectrum, elucidating the stages encompassed in astute decision-making.

The recommended stages for optimal decision-making are:

1. Pinpointing personal predispositions that might jeopardize flight safety.
2. Adopting techniques to modify behavior.
3. Understanding and managing stress effectively.
4. Cultivating skills to assess risks.
5. Mobilizing all available resources.
6. Appraising the proficiency of one's aeronautical decision-making capabilities.

Managing Risks in Aviation

Risk management aims to pinpoint safety-related hazards and alleviate the corresponding dangers actively. It is a pivotal part of aeronautical decision-making. A pilot who adheres to sound decision-making protocols can significantly diminish or even eradicate the inherent risks associated with a flight. The foundation for making informed decisions lies in firsthand or secondhand experiences complemented by education.

Let's use the wearing of bicycle helmets as an example. Within a brief period, wearing bicycle helmets has become a standard safety protocol, rendering those who need to wear them outliers. However, these individuals might recognize the importance of wearing a helmet through firsthand or secondhand experiences. For instance, a direct realization might stem from a cyclist getting into an accident without a helmet, resulting in an injury. A secondhand realization might arise when someone they know suffers an injury from a biking accident due to the absence of a helmet.

Navigating through the aeronautical decision-making process, it is vital to anchor decisions to the four core tenets of risk management:

- Avoid Unwarranted Risks. While flying inherently carries risks, unnecessary risks do not yield proportional benefits.
- Delegate Risk Decisions Judiciously. Decisions concerning risks should be the prerogative of individuals capable of both formulating and executing risk control measures.
- Assume Risks Judiciously. Only accept risks if the advantages outweigh the potential hazards.
- Weave Risk Management into All Planning Phases. Given that risks are an integral part of every flight, ensuring safety mandates the consistent integration of effective risk management, spanning from preflight preparations to all subsequent flight phases.

While lapses in everyday decision-making might not always culminate in severe consequences, the aviation sphere offers limited room for errors. As aeronautical decision-making bolsters the navigation of the aviation milieu, pilots must acquaint themselves with and consistently employ aeronautical decision-making strategies.

Understanding Hazards and Risks

Critical components of Aeronautical Decision-Making include understanding hazards and assessing associated risks. A hazard can be described as any genuine or perceived situation, event, or condition encountered by a pilot. Once confronted with a hazard, the pilot evaluates it, considering multiple factors. The pilot then gauges the potential consequences of that hazard, leading to the determination of its risk. Thus, risk can be understood as a pilot's evaluation of the immediate or combined threats a hazard poses. It is worth noting that different pilots perceive and evaluate hazards differently.

The Role of Attitudes in Decision Making

A pilot's capability to fly safely is more than just their physical state or recent flight experience. Attitudes, or the inherent tendencies to react to certain situations, people, or events in a specific way, play a pivotal role in decision-making. Research pinpoints five prevalent hazardous attitudes that could compromise sound decision-making and proper authority usage: anti-authority, impulsivity, invulnerability, macho, and resignation.

Examples of Hazardous Attitudes:

- **Anti-authority**: Disregarding established guidelines or protocols. For instance, a pilot choosing to ignore air traffic control instructions because they believe they know better. The antidote to this attitude would be reminding oneself, "It's important to follow the rules; they are usually based on someone else's past experiences."
- **Impulsivity**: Making hasty decisions without fully thinking them through. A pilot might decide to take off without a thorough pre-flight check, believing they have done it enough times to skip it. The antidote to this attitude would be reminding oneself, "It's important to follow the rules; they are usually based on someone else's past experiences."
- **Invulnerability**: Believing that dangers apply to others, but not to oneself. For example, a pilot thinking, "Accidents happen, but they won't happen to me." The antidote for this attitude could be: "It can happen to me."
- **Macho**: Taking unnecessary risks to show off or prove oneself. A pilot might perform a tricky maneuver close to the ground to impress others, despite the obvious dangers. The appropriate response here might be: "Taking unnecessary risks is foolish."
- **Resignation**: Feeling powerless or thinking that one's actions will not make a difference. For instance, a pilot facing worsening weather conditions might think, "Whatever is going to happen will happen," instead of seeking a safer alternative. To counteract resignation, a pilot can tell themselves, "I can make a difference; my actions matter."

Evaluating Risk

For solo pilots, the process of evaluating risk can be complex. Often, they are their checks and balances when making decisions. Consider a scenario where a pilot, after a strenuous 16-hour flight, must judge if fatigue has set in. The pilot might underestimate the exhaustion, thinking they can carry on. Objectives drive many pilots, and when presented with a mission, there might be a tendency to overlook personal constraints, giving undue emphasis to unrelated matters. For instance, emergency medical service helicopter pilots, more than others, have occasionally been observed making decisions where they place excessive priority on a patient's well-being. Such pilots might undervalue tangible risks like exhaustion or adverse weather, emphasizing the emotional pull of the situation. This inclination to focus on emotional or intangible factors can lead solo pilots into precarious situations, making them potentially more susceptible to risk than a team of pilots.

Addressing Risk

Understanding the potential risks is just the beginning. One effective tool solo pilots employ to address risk is the IMSAFE checklist, ensuring they are both mentally and physically prepared to fly:

- **Illness** - Is my health compromised in any way? Being unwell can be a clear risk to piloting capabilities.
- **Medication** - Have I consumed any drugs that could influence my cognitive abilities or induce drowsiness?
- **Stress** - Am I grappling with external pressures, be it work-related or personal issues like finances, health, or familial concerns? Though regulations might not explicitly mention stress as a grounding factor, it is vital to understand its potential to divert attention and hamper performance.
- **Alcohol**: Have I had any in the last eight hours? How about 24 hours? Even tiny amounts, be it a shot of spirits, a glass of beer, or a serving of wine, can negatively affect flying capabilities. Moreover, alcohol can increase a pilot's chances of disorientation and hypoxia.
- **Fatigue**: Do I feel rested or exhausted? One of the subtle yet potent threats to flying safety is fatigue, which may not manifest until grave mistakes occur.
- **Emotion** - Am I in an emotional turmoil or distress?

The PAVE Framework

An essential method to recognize and manage potential hazards is the PAVE framework. This tool offers a systematic approach for pilots to break down flight risks into four clear categories: Pilot's capacity (P), Aircraft's condition (A), Environment (C), and External influencers (E). This checklist becomes an intrinsic component of a pilot's decision-making regimen.

The PAVE framework offers pilots a structured method to scrutinize each risk element before every flight. After pinpointing the risks associated with a flight, the pilot must determine if these risks, either singularly or collectively, are manageable. If deemed too risky, the pilot should be prepared to postpone or cancel the flight. On the other hand, risk-mitigating strategies should be established if the decision is to proceed. One approach to maintain safety is for pilots to develop personal benchmarks for each risk category, contingent upon their current expertise and skillset.

P = Pilot's Capacity

- The pilot remains a pivotal variable in any flight's risk profile. Essential questions include:
- "Am I adequately equipped for this journey?" in terms of skill level, recent flying experiences, emotional and physical state.
- The IMSAFE checklist remains a valuable tool for this assessment.

A = Aircraft's Condition

The aircraft's state can pose restrictions or challenges. Key considerations include:

- Is this aircraft suitable for my intended journey?
- Am I well-versed with and have recent experience flying this model?
- Can the aircraft manage the anticipated cargo and passenger weight?

V = Environment

The elements play a crucial role when flying:

- Weather: Personal benchmarks concerning weather are crucial. As pilots evaluate meteorological conditions, ponder the following:
  - Are the current visibility and cloud cover within safe limits?
  - Remember, actual weather conditions can deviate from predictions.
  - Are stormy conditions in the forecast or currently observed?
  - If clouds are present, is there a risk of icing? How do the altitude's current temperature and the temperature/dew point gap measure up?
- Landscape: It is imperative to assess the geography of the planned route.
- Airspace: Always verify the status of the airspace and stay alert for any temporary flight restrictions.

E = External Influencers

These are the external aspects compelling pilots to proceed with a flight, potentially overshadowing safety concerns. Such pressures could stem from:

- The urge to highlight one's piloting competencies.
- The impulse to astonish an observer. (Aviation’s most perilous phrase might be "See this!")
- The innate determination to achieve set goals.
- Emotional strains, especially when one must admit that they might not be as adept or experienced as desired. Ego can indeed be a compelling external influencer!

Chapter 9 - Decision-Making in Aviation

Grasping the essence of the decision-making process equips pilots with the foundational tools needed for advanced aeronautical decision-making and Single-pilot Resource Management techniques. While scenarios, like an engine malfunction, demand an instantaneous pilot reaction based on set protocols, in-flight changes typically allow pilots to process the situation, gather pertinent data, and weigh risks before finalizing a decision.

Both risk management and risk intervention surpass mere terminological understanding. These terms denote a holistic decision-making approach tailored to systematically spot potential threats, evaluate associated risks, and chart the optimal action plan. This method involves spotting hazards, gauging the inherent dangers, scrutinizing the available control measures, deciding on control implementation, executing these controls, and overseeing the outcomes.

The 3P Model: Perceive, Process, Perform

The 3P model for aeronautical decision-making offers a systematic and intuitive approach applicable during all flight stages. When employing this model, a pilot should:

- Perceive: Recognize the circumstances surrounding a flight.
- Process: Evaluate how these circumstances might affect flight safety.
- Perform: Carry out the best course of action.

Pilots should continuously employ the Perceive, Process, Perform, and Evaluate approach for every flight-related decision. While human error is inevitable, consistently applying this model to recognize and mitigate potential risks enhances flight safety and effectiveness.

For risk management, depending on the specific flight activity and the time at hand, the processing might occur in one of three distinct periods. Notably, most flight training occurs within the "immediate" risk management period.

Decision-Making Amidst Fluid Circumstances

Utilizing structured models provides a comprehensive approach to decision-making. Effective decisions arise when pilots accumulate all pertinent data, scrutinize it, dissect the choices, rank these options, choose a path forward, and subsequently assess that chosen path for its viability.

However, in specific scenarios, the luxury of time is not on the pilot's side to employ these analytical decision-making methods. Consider a chef in a bustling kitchen, for instance. While they might start with a particular recipe or plan in mind, the dynamic environment of the kitchen, varying demands, and changing situations require them to adapt and make snap decisions. This instinctive, on-the-spot form of decision-making is termed automatic or naturalized decision-making.

Intuitive Decision-Making

Professionals grappling with uncertain tasks in time-sensitive situations tend to rely on familiarity. Instead of painstakingly weighing the advantages and disadvantages of various alternatives, they rapidly visualize the potential outcomes of a select few responses to the situation. These experts frequently choose the first workable answer that comes to mind. While this might only sometimes be optimal, it frequently leads to notably effective results.

This decision-making mode has been labeled "intuitive" and "reflexive decision-making." Whether it is a firefighter evaluating the best way to tackle a blaze or a chess master strategizing a move, this approach depends on recognizing patterns and consistencies illuminating options in multifaceted scenarios. Experts often navigate situations by initiating actions based on past experiences, which pave the way for innovative adjustments.

Rooted in extensive training and experience, this reflexive approach to decision-making comes to the fore, especially during emergencies when deliberative analysis is not feasible. As one gains more experience and training, one's proficiency in intuitive decision-making is enhanced. Over time, a pilot, for instance, will seamlessly integrate various decision-making techniques, aligning them with their training and experiences.

Leveraging Available Tools

For sound decision-making during flight operations, pilots must be acutely aware of available resources. Identifying these valuable tools and information sources is a cornerstone of aeronautical decision-making education. However, more than merely recognizing resources is required. Pilots must also hone the skills to assess when and how to use these resources and gauge the implications of their use on flight safety.

Situational Awareness

Situational awareness refers to comprehending and acknowledging all aspects within the core five risk components (flight dynamics, the pilot, the aircraft's state, environmental conditions, and operational nature) that shape any aviation scenario. This understanding influences safety measures taken before, throughout, and after the flight. To maintain optimal situational awareness, a pilot must grasp the importance of every factor involved in aviation and anticipate their potential outcomes. A pilot must look at the big picture rather than just one specific aspect.

Beyond knowing the aircraft's physical location, a pilot should be attuned to the unfolding circumstances around them.

Factors like fatigue, stress, and excessive tasks can make pilots overly focused on a single perceived crucial detail, diminishing their comprehensive understanding of the flight. Often, disruptions that pull a pilot's focus away from the primary task can contribute to mishaps.

Managing Tasks Efficiently

Ensuring tasks are efficiently completed involves strategizing, ranking, and organizing tasks to prevent becoming overwhelmed. As pilots garner more experience, they become adept at forecasting future task demands and can utilize periods of low activity to prepare for more demanding phases.

Moreover, pilots should tune in to Automated Terminal Information Service (ATIS), Automated Surface Observing System (ASOS), or Automated Weather Observing System (AWOS) when accessible and subsequently keep an ear on the tower frequency or Common Traffic Advisory Frequency (CTAF) to gain insights into anticipated traffic conditions. Identifying when one is overwhelmed is integral to task management.

The initial symptom of an overburdened workload is that, despite the increased effort, the pilot's productivity drops. With a growing list of tasks, a pilot's attention becomes fragmented, leading them to hone in on only one activity. When overwhelmed, a pilot might lose sight of various inputs, leading to decisions based on incomplete data and elevating the risk of mistakes.

In such demanding situations, a pilot must pause, reflect, recalibrate, and set priorities. Grasping how to reduce workload becomes essential.

Chapter 10 - Airfield Operations

Any specified place, whether on land or in the water, intended for the arrival or departure of aircraft is referred to as an airfield.. This category includes specialized facilities such as seaplane bases, helipads, and venues suitable for tilt rotorcrafts. It encompasses the spaces earmarked for airfield structures, facilities, and the associated infrastructure.

Typically, airfields are divided into two categories: those that have control towers and those that do not. These can be further categorized into:

- Public Airfields - Accessible to the general populace.
- Government/Military Airfields - Managed by the National Aeronautics and Space Administration, the military, or other federal organizations.
- Private Airfields - Exclusively for private or limited access, not available for public use.

Controlled Airports

An airport with an active control tower falls into this category. Air traffic control takes on the responsibility of ensuring that air traffic flows safely, systematically, and swiftly. Such services become essential at airports where the nature and/or volume of air traffic demands it.

Uncontrolled Airports

Uncontrolled airports function without an active control tower. While two-way radio communication is not mandatory here, it is advantageous for pilots to stay informed by tuning into the designated frequency; this promotes awareness among other nearby aircraft. The essence of managing traffic at such airports is pinpointing the right shared frequency. This frequency is often called the Common Traffic Advisory Frequency. A Common Traffic Advisory Frequency is allocated for communication to provide advisories while operating around an airport without an active control tower. This frequency could be tied to a Universal Integrated Community (UNICOM), MULTICOM, FSS, or a tower highlighted in relevant aviation documents. UNICOM is a non-official air-to-ground radio channel that might offer airport details at public access airports without a tower or an FSS.

Pilots always enter the traffic patterns at the designated pattern altitude for uncontrolled airports. The way one enters the pattern depends on how they approach it. For example, if arriving from the downwind side, the favored entry method is to align with the pattern at a 45-degree angle to the downwind leg, merging with the pattern around its midpoint.

Resources for Airport Information

For remote pilots flying near an airport, accessing up-to-date data is crucial. This information offers insights like communication channels, out-of-service runways, or ongoing construction at the airport. Key information repositories are:

- Aviation Maps
- Notices to Airmen (NOTAMs)
- Automated Terminal Information Service (ATIS)

We talked about NOTAMs in previous chapters. Let us focus on the other two information repositories.

Aviation Maps

Avion maps are akin to a driver's road map. They equip remote pilots with data about the zones where they plan to fly. The two primary aviation maps for pilots are Regional Maps and VFR Urban Area Maps.

Regional Maps are designed to cover specific regions, allowing pilots to understand the nuances of that area. They highlight airspace configurations, airports, notable landmarks, and other crucial navigation points within that region.

VFR Urban Area Maps are utilized by pilots flying under visual conditions. The Urban Area Maps specifically target densely populated regions, providing a more detailed view of city landscapes. This granularity is especially beneficial for drone operators who might be conducting operations in urban settings, making them aware of obstacles, notable buildings, and other urban-specific navigation aids.

For a comprehensive list of aviation maps, related publications, pricing details, and order instructions, visit the FAA's Aeronautical Navigation Products portal at: www.aeronav.faa.gov.

Automated Terminal Information Service (ATIS)

On a specific frequency, the Automated Terminal Information Service broadcasts the most recent local meteorological conditions as well as other crucial non-control information continuously. While updates are hourly, more frequent updates can occur if local conditions shift significantly. ATIS relays vital data such as weather conditions, active runways, and specific air traffic control procedures. Each airport with an ATIS service will have a designated radio frequency to access the broadcast. This frequency is typically separate from the central air traffic control channels to avoid congestion.

Each Automated Terminal Information Service recording is identified by a unique code, which changes with every update. For instance, Automated Terminal Information Service Bravo will succeed Automated Terminal Information Service Alpha, followed by Automated Terminal Information Service Charlie, Automated Terminal Information Service Delta, and so on through the phonetic alphabet.

With technological advancements, many airports now offer Digital ATIS, where the information is transmitted digitally, enabling it to be displayed directly on cockpit instruments or flight management systems.

While ATIS is geared toward manned aviation, understanding its functionality can benefit drone operators, especially those operating near airports. Knowing current airport conditions, weather, and other crucial details can enhance situational awareness and promote safer operations.

Geographical Coordinates (Longitudinal and Latitudinal Lines)

The equator is a theoretical line equidistant from Earth's poles. Circles running parallel to the equator, moving east-west, are referred to as latitudinal lines. These lines are used to calculate latitude measurements that are either north (N) or south (S) of the equator. The angular span from the equator to any pole covers a quarter of a circle, which is 90°. Between latitudes 25° and 49° N are the contiguous 48 states of the United States.

Meridians, also known as longitude lines, are a right angle to the equator and run from the North Pole to the South Pole. The "Prime Meridian" traverses Greenwich, England, and serves as the foundational line for measuring degrees either east (E) or west (W) up to 180°. Between longitudes 67° and 125° W, the contiguous 48 states of the United States are located. These longitudinal lines are indicated in Figure 11-3 by the "Longitude" arrows.

By using the longitudinal and latitudinal coordinates of a site, it is possible to locate any precise geographic position. For instance, New York City is situated at 40° N latitude and 74° W longitude, while Los Angeles lies around 34° N latitude and 118° W longitude.

Magnetic Deviation

The angle between magnetic north and true north (TN) is known as magnetic deviation (MN). This angle is described as an east or west deviation, depending on whether MN lies to the east or west of TN.

The magnetic North Pole is approximately 71° N latitude and 96° W longitude, 1,300 miles distant from the actual geographic North Pole. If Earth had a consistent magnetic field, a compass needle would invariably point toward the magnetic pole. Under such circumstances, the deviation between TN (represented by geographical lines) and MN (indicated by magnetic lines) could be gauged wherever these lines intersect.

However, Earth's magnetism is only sometimes consistent everywhere. Within the United States, while a compass typically points towards the general direction of the magnetic pole, it can differ significantly in specific regions. As a result, the precise deviation at numerous locations has been meticulously charted. This direction and magnitude of deviation, which undergoes slight changes periodically, are illustrated on many aviation maps as dashed magenta lines—these lines, known as isogonic lines, link points with the same magnetic deviation. Conversely, the line that joins points without deviation between TN and MN is termed the agonic line. The intricate bends and shifts in these lines are influenced by distinct geological factors that affect the magnetic forces in those regions.

Tower Hazards

Pilots should exercise utmost caution when flying at altitudes below 2,000 feet AGL due to numerous tall structures, like radio and TV antenna towers. Many of these towers rise above 1,000 feet AGL, with some even surpassing the 2,000-foot mark. A significant danger associated with these structures is their supporting guy wires, which, even in clear weather, can be challenging to spot. These wires can become invisible during twilight or in decreased visibility conditions. Given that these wires can stretch approximately 1,500 feet from the main structure, it is advisable to maintain a horizontal distance of at least 2,000 feet from any such tower.

Moreover, pilots should be aware that recently erected towers might not be represented on their current charts, especially if the data was unavailable at the time of chart printing.

Chapter 11 - Sectional Charts

Sectional Charts are the predominant maps utilized by pilots today. These maps feature a scale of 1:500,000, where 1 inch equates to 6.8 statute miles, or 86 nautical miles (SM) This scale permits the inclusion of in-depth details on the map.

These maps offer many details, encompassing airport specifics, navigational aids, airspace classifications, and geographical contours. Pilots can decipher most of its content by consulting the map's legend. It is also essential for pilots to review other legend details on the map, as it contains crucial data like air traffic control frequencies and comprehensive airspace knowledge. These maps undergo updates every six months. However, certain regions outside the contiguous United States may only see annual revisions.

Historical Context of Sectional Charts

The dawn of aviation brought about an urgent need for standardized navigational aids. In the early 20th century, as airplanes began to rise in popularity and practicality, pilots primarily relied on visual landmarks for navigation, such as roads, railroads, and natural formations. While this was feasible for short distances, longer journeys posed a challenge.

As airmail routes expanded during the 1920s, the need for better navigation tools became evident. The first aerial charts were created to aid airmail pilots. These charts highlighted key landmarks, beacon light locations, and essential terrain details. However, as aviation continued to expand, the limitations of these early charts became apparent. With the increasing complexity of air traffic, especially around growing metropolitan areas, there was a need for more detailed navigational information. This led to the development of sectional charts.

Introduced by the U.S. government in the 1930s, sectional charts brought a new level of detail to pilot navigation. They offered an accurate representation of the ground below, including topographical information, significant landmarks, airports, and, later, specific airspace classifications. They were named "sectional" because the entire country was divided into sections, with each chart detailing a particular part of the nation. This allowed pilots to use a series of charts for cross-country flights, ensuring they always had detailed information about their current location.

Over the years, as aviation requirements changed, these charts' layout and information density changed as well. With new navigational aids, airspaces, and regulations, sectional charts were continually updated to remain a pilot's primary tool for visual navigation. Today, while technology has given rise to GPS and digital navigation tools, sectional charts remain an essential tool and reference for pilots, both for their historical significance and their practical use in understanding the complexities of airspace. Furthermore, they are a crucial part of the FAA part 107 exam. Therefore, every aspiring exam taker should be familiarized with them.

Decoding Airport Data

Airport data on sectional charts provides pilots with essential information about an airport's facilities and services. Interpreting this data correctly is crucial for safe and efficient operations, especially when landing at unfamiliar airports. Here is an overview to decoding airport data on sectional charts:

- **Airport Symbol**: The symbol used for an airport on the chart indicates the type of runway and whether the airport has a control tower.
    - Solid Blue Circle: Paved runway with a control tower.
    - Solid Magenta Circle: Paved runway without a control tower.
    - Solid Blue Airport Layout: Detailed runway layout indicating a paved runway with a control tower.
    - Solid Magenta Airport Layout: Detailed runway layout indicating a paved runway without a control tower.
    - Outlined Airport Layout (either blue or magenta): Unpaved runway, with the color indicating the presence (blue) or absence (magenta) of a control tower.
- **Runway Length**: The longest runway's length is depicted in hundreds of feet. For example, "27" would indicate a runway length of 2,700 feet.
- **Control Tower Frequency**: If an airport has a control tower, the tower's frequency is listed next to the airport symbol.
- **Other Frequencies**: Frequencies for other services like ground control, Automatic Terminal Information Service (ATIS), and Flight Service Stations (FSS) might also be listed.
- **Lighting**: An airport symbol with a star (*) indicates that the airport has lighting, and it is essential for nighttime operations.
- **Elevation**: The elevation of the airport (in feet above mean sea level) is provided, which is crucial for calculating aircraft performance, especially during takeoff and landing.
- **Beacon**: A star in a circle indicates the presence of a rotating airport beacon, usually operating from sunset to sunrise.
- **Services & Facilities**: Some charts will use specific symbols or abbreviations to indicate available services like fuel, repairs, or customs. These are typically found in the airport's directory section.
- **Other Data**:
    - Right Traffic: If an airport has a traffic pattern that requires right-hand turns for one or more runways, it will be indicated with an "R" and the runway number. Most traffic patterns are standard left-hand turns.
    - Obstructions: Nearby obstructions like tall towers or other hazards might be noted.

- Special Procedures: Some airports have special procedures due to noise abatement, nearby airspace, or other reasons. These might be referenced on the chart, but pilots should consult the Airport/Facility Directory (A/FD) for detailed information.

Special Use Airspaces

In the previous chapters we have already discussed Special Use Airspaces. In this section we point out how they appear on a sectional chart.

- **Restricted Areas (R):** Surrounded by blue hashed lines and identified by a "R" and a number (e.g., R-4401).
- **Prohibited Areas (P):** Outlined with blue hashed lines like restricted areas but labeled with a "P" followed by a number (e.g., P-56).
- **Warning Areas (W):** Outlined with blue hashed lines like restricted areas but labeled with a "W" followed by a number (e.g., W-237).
- **Military Operation Areas (MOAs):** Outlined with magenta hashed lines and labeled with the MOA's name.
- **Alert Areas (A):** Outlined with magenta hashed lines and labeled with an "A" followed by a number (e.g., A-211).
- **Controlled Firing Areas (CFAs):** Not depicted on sectional charts because the activities pose no hazard to aircraft.
- **National Security Areas (NSAs):** Depicted on sectional charts with a broken magenta outline and labeled with the NSA's name.

Chart Symbols for Services

Sectional charts include various symbols to denote the availability of services at or near airports. These symbols can be particularly useful for pilots planning cross-country flights or looking for specific services. Here is a summary of the most common chart symbols for services:

- **Fuel**: Depicted by a small solid circle. It indicates that aviation fuel is available at the airport.
- **Repairs**: Symbolized by an open wrench icon. This denotes that aircraft repair services are available on the field.
- **Airport Beacon:** A star inside a circle symbolizes the presence of a rotating airport beacon, which is usually operational from sunset to sunrise.
- **Controlled Field:** Airports with a control tower are denoted with blue airport symbols (either solid blue or blue outlines). The neighboring listing will provide the control tower frequency.
- **Uncontrolled Field:** Airports without a control tower are marked with magenta symbols. For pilot communication, the airport's UNICOM frequency or Common Traffic Advisory Frequency (CTAF) will be shown nearby.

- **Landing Fees:** While sectional charts do not typically indicate landing fees, pilots often consult Airport/Facility Directories or other resources for this information.
- **Customs**: An airport with customs services for international flights is denoted by a letter "C" in a circle.
- **Military Airports**: Depicted by symbols using two or more runways intersecting in a circle. It indicates that the field is used for military purposes, and civilian aircraft might need special clearance to land or overfly.
- **Parachute Jumping Area**: A parachute symbol denotes an area where parachute jumping activity takes place. Pilots should exercise caution and check NOTAMs for activity times.
- **Camping**: Some charts may use a tent symbol to indicate airports where camping is available, particularly useful for fly-in camping trips.
- **Seaplane Base:** Depicted by an anchor symbol, indicating a location where seaplanes can land and take off.
- **Heliport**: A circle with an "H" in the middle represents a heliport.
- **Weather Information:** An enclosed "W" symbol denotes that Automated Weather Observing Systems (AWOS), Automated Surface Observing Systems (ASOS), or other weather information services are available.

Visual Checkpoints

Visual checkpoints are landmarks that pilots can use for visual navigation. These checkpoints are easily recognizable from the air and serve as references to confirm an aircraft's position relative to the ground. They are particularly valuable when pilots are navigating in congested areas, especially near busy airports, where precise positioning can be crucial for safe operations.

Types of Visual Checkpoints:

- **Natural Landmarks**: These can include mountains, rivers, lakes, coastlines, and large forests.
- **Man-made Landmarks**: These encompass structures like bridges, dams, tall buildings, factories, stadiums, and distinct road intersections.
- **Airport Landmarks**: Some airports might have unique structures, colored roofs, or distinct taxiway patterns that make them easily recognizable from the air.

Visual checkpoints are marked with a magenta flag symbol on sectional charts. Alongside the flag, there is typically a label describing the checkpoint. For instance, a bridge might be labeled as "Suspension Bridge," or a stadium might be identified by its name. Some charts might also provide additional information about the checkpoint, like its elevation or a brief description.

Topographical Information

Pilots can see a visual depiction of the topography and physical characteristics of the Earth thanks to the topographic information on sectional charts.

Contours:

- Contour Lines: These are curved lines that link locations with the same elevation. They give a 2D representation of the terrain's shape and elevation.
- Interval: The contour interval is the height difference between two adjacent contour lines. On sectional charts, this interval is typically set to represent a specific elevation change, such as 500 feet.
- Closely Spaced Lines: Indicate a cliff or mountain ridge to denote a steep area.
- Widely Spaced Lines: Indicate flatter terrain.

Colors:

- Green: Represents areas of lower elevation, typically below 2000 feet Mean Sea Level.
- Light Brown: Represents higher elevations, typically above 2000 feet MSL.

Maximum Elevation Figures (MEFs):

These figures are found inside quadrants on the chart and show the highest elevation in that quadrant, including terrain and other obstructions. They provide a quick reference for safe flying altitudes. MEFs represent either the elevation of the highest known feature (terrain, trees, structures, etc.) in each quadrant or 100 feet above the highest obstacle for areas dominated by structures. The Maximum Elevation Figure is displayed as a large, bold number in the center of each quadrangle.

By flying at or above the MEF for a given area, pilots can be confident that they are clear of all obstacles and terrain in that quadrangle. However, it is always crucial for pilots to be vigilant and not solely rely on the MEF for obstacle clearance, especially when flying at low altitudes.

Interpreting Chart Marginal Information

Marginal information on sectional charts provides essential details that are crucial for pilots to understand and utilize the chart effectively. A sectional chart's margins are filled with useful reference material.

- **Chart Name and Edition**: At the top center, you will find the chart's name (e.g., "Denver Sectional Chart") and its current edition number or date. Ensuring you are using the most recent chart edition is essential for safety.

- **Scale**: As previously stated, sectional charts frequently have a scale of 1:500,000, meaning that 1 inch on the chart corresponds to 500,000 inches on the ground. The scale can help pilots estimate distances and flight times.
- **Legend**: The legend decodes the various symbols used on the chart, from airport data to topographical features. Familiarizing yourself with the legend is vital to understand the chart's content fully. The full sectional chart legend is available in the "DIGITAL RESOURCES" section at the end of the guide.
- **Frequency Information:** This section lists communication frequencies for Flight Service Stations, Automatic Terminal Information Service, Weather, and other essential services.
- **Adjacent Charts**: There's typically a diagram showing the adjacent sectional charts, helping pilots determine which chart they need if they are crossing over into another chart's coverage area.
- **Conversion Tables:** Useful for converting nautical miles to statute miles, calculating the time required to travel a specific distance at a given speed, and more.

Interpretation Checklist

One of the most difficult parts of the FAA Part 107 exam is interpreting sectional charts. This checklist provides a structured approach to this problem, but always remember that with experience, some of these steps will become second nature. Until then, methodically working through the chart using the checklist can ensure thorough understanding and a successful exam.

Preliminary Steps:

- Ensure the sectional chart is the latest edition.
- Identify the chart's name and coverage area (found in the top margin).
- Familiarize yourself with the scale of the chart (typically 1:500,000).

Legend and Symbols:

- Refer to the chart's legend to understand symbols, lines, and colors.
- Identify the various airport symbols, noting the differences between controlled, uncontrolled, and private airports.
- Recognize symbols for diverse types of navigational aids, such as VORs.

Airspace and Boundaries:

- Determine the various airspace classifications (Class B, C, D, etc.).
- Look for boundaries of special use airspace (MOAs, Restricted Areas, etc.).
- Identify Transition Areas, typically depicted by a thin blue or magenta line.

Topographical and Cultural Features:

- Check for contour lines to understand terrain elevation and relief.
- Recognize significant bodies of water, roads, railways, and populated areas.
- Look for specific spot elevations, especially on mountain peaks.

Airport Data:

- For each airport, determine runway length, surface type, and lighting availability.
- Determine the frequencies, control towers, and other pertinent services.

Navigational Aids:

- Locate VORs, NDBs, and other navaids.
- Check for associated frequencies and Morse code identifiers.
- Identify the compass rose around VORs for radial navigation.

Obstructions and Safety:

- Scan for tall towers, antennas, or other obstructions.
- Check for the Maximum Elevation Figures (MEFs) in each quadrant.
- Look for any caution notes or special warnings.

Special Features and Points:

- Identify any VFR checkpoints, which can be landmarks or intersections.
- Look for parachute jumping areas, glider operations, or other unique activities.

Communication Information:

- Refer to the margins for frequencies related to Flight Service Stations, weather reports, and Automated Terminal Information Service.
- Locate the contact details for nearby ARTCC and sector frequencies.

Insets and Special Zones:

- If the chart has insets (typically around major cities), ensure you review them for additional detail.
- Recognize any TFRs (Temporary Flight Restrictions) if marked.

Review:

- Recheck the entire chart to make sure nothing was overlooked.
- Consider using a highlighter or marking tool to emphasize critical areas or routes for

Legend and Bonus Flashcards

We understand the importance of visual aids in enhancing comprehension and retention, especially when it comes to aviation and navigation topics. However, due to copyright restrictions, we were unable to include images of sectional charts directly within the pages of this book.

But don't be alarmed! We've got your back. To ensure that you still benefit from hands-on exercises and a complete learning experience, we have incorporated 50 sectional chart exercises in a digital format. You can easily access and download these resources, which are tailored to complement and reinforce the content discussed in the book.

For your convenience, all these digital sectional chart flashcards are housed in the DIGITAL BONUSES section of the book.

Conclusion

Having delved deep into the FAA Part 107 and its nuances, it is now time to apply what you have learned. This book was created to give you the information and understanding you'll need for your drone operations and, most crucially, for your upcoming exam.

Remember, while the details and regulations might seem overwhelming at first, they are the foundation of safe and effective drone flight. With the understanding you have gained from this guide, you are now better prepared to tackle the challenges of the exam and the real-world scenarios that will follow.

The drone business is going through an exciting and dynamic period. As you move forward, continually seek knowledge, remain vigilant, and always prioritize safety.

Wishing you luck on the FAA Part 107 exam Success is attainable with commitment and planning. Here is to your future as a certified remote pilot!

Practice Test

1. Which class of controlled airspace typically has its configuration visualized as an inverted wedding cake?

A) Class A
B) Class B
C) Class C
D) Class G

2. You intend to operate your drone in a particular airspace. The elevation begins at 1,200 feet above the ground level (AGL) and rises to 18,000 feet MSL but does not include that height. Which class of airspace will you be operating in?

A) Class A
B) Class C
C) Class D
D) Class E

3. If a remote pilot wishes to fly in an airspace that starts from the surface and extends upward until it reaches the beginning of the overlying Class E airspace, which class of airspace is the pilot intending to fly in?

A) Class A
B) Class D
C) Class G
D) Class E

4. Which type of Special Use Airspace has operations that will cease if a non-participating aircraft approaches, and is not marked on charts?

A) Prohibited Areas
B) Alert Areas
C) Military Operation Areas
D) Controlled Firing Areas

5. You come across a Special Use Airspace on a chart with the designation "W-285." What type of airspace is this, and what should you be cautious of?

A) Warning Area; activities potentially dangerous to non-participating aircraft.
B) Restricted Area; unusual and potentially unseen dangers.
C) Prohibited Area; area where flight is entirely disallowed.
D) Military Operation Area; designated for military training exercises.

6. In which Special Use Airspace are aircraft movements strictly disallowed due to security considerations or factors vital to national interests?

A) Alert Areas

B) Military Operation Areas

C) Restricted Areas

D) Prohibited Areas

7. What is the primary purpose of Local Airport Advisory (LAA) and how does it operate?

A) To control the air traffic around airports and operates through radio communication.

B) To provide guidance about the local airport and weather updates via a specific ground-to-air channel.

C) To establish secure conditions for space agency missions and operates through satellite communication.

D) To manage Military Training Routes and operates through Automated Weather Observing Stations.

8. What distinguishes the numerical identifiers of Military Training Routes (MTRs) below and above 1,500 feet AGL?

A) Below 1,500 feet AGL use a three-digit code, above use a four-digit code.

B) Below 1,500 feet AGL use a four-digit code, above use a three-digit code.

C) Both use a four-digit code, but with different prefixes.

D) Both use a three-digit code, but with different prefixes.

9. For which of the following reasons might Temporary Flight Restrictions (TFR) be established?

A) To facilitate airshows and exhibition flights.

B) To safeguard national disaster zones for compassionate reasons.

C) To provide a clear pathway for migratory bird patterns.

D) To control traffic around National Security Areas (NSAs).

10. How can pilots access Notices to Airmen (NOTAMs)?

A) Solely through the National Airmen Bulletin System.

B) Via subscriptions from the Document Oversight Office or digitally through platforms like PilotWeb.

C) Only through automated systems at local airports.

D) Exclusively via radio communications with air traffic control.

11.What does METAR KLAX 141930Z AUTO 15018G24KT 2SM -RA FG SCT006 OVC010CB 16/15 A2985 RMK PRESRR indicate about the wind speed and direction at Los Angeles International Airport?

A) Wind coming from 150° at 18 knots, gusting to 24 knots

B) Wind coming from 180° at 15 knots, gusting to 24 knots

C) Wind coming from 240° at 15 knots, gusting to 18 knots
D) Wind coming from 150° at 24 knots, no gusts reported

12. In a METAR report, what does "AUTO" signify?
A) The report is automated
B) The report is automatically corrected
C) The report has an automatic update frequency
D) The report includes automatic pressure readings

13. What is the visibility reported in METAR KLAX 141930Z AUTO 15018G24KT 2SM -RA FG SCT006 OVC010CB 16/15 A2985 RMK PRESRR?
A) 1 statute mile
B) 2 statute miles
C) 6 statute miles
D) 10 statute miles

14. A TAF report for JFK Airport, beginning with TAF KJFK 121400Z 1214/1314, indicates:
A) The report was issued on the 12th day of the month at 1400Z
B) The forecast is valid from 1200Z on the 14th to 1400Z on the 13th
C) The report covers a period of 48 hours
D) The forecast is valid for a 12-hour period

15. What type of weather condition is described by the descriptor "TSRA" in a TAF report?
A) Thunderstorm with light rain
B) Thunderstorm with heavy rain
C) Thunderstorm with rain
D) Thunderstorm with mist

16. In an AIRMET, what does "TANGO" refer to?
A) Icing conditions
B) Turbulence
C) Thunderstorms
D) Tornadoes

17. Convective SIGMETs are issued primarily for which conditions?
A) Light turbulence and icing
B) Severe thunderstorms with wind speeds over 50 knots
C) Cloud cover over 40% of an area
D) Hail sizes less than ¾ inch

18. Which of the following is the correct interpretation for an area affected by turbulence described as "FROM 30ESE TBE TO INK TO ELP"?

A) The turbulence affects the entire region between the three mentioned waypoints.
B) The turbulence starts 30 nautical miles east-southeast of TBE and ends at ELP.
C) The turbulence affects only the direct route between TBE and ELP.
D) The turbulence affects areas within 30 nautical miles of TBE and ELP.

19. Which aviation weather report provides details about the atmospheric conditions within a five-statute mile radius of an airport?
A) METAR
B) Convective SIGMET
C) TAF
D) AIRMET

20. In a Convective SIGMET, what does the term "PROB30" indicate?
A) 30% chance of precipitation in the area
B) 30% chance of thunderstorms and rain within the forecast
C) 30% humidity level in the region
D) 30% likelihood of clear skies

21.Which of the following best sums up how air density and temperature relate when there is a constant pressure?
A) Directly proportional
B) Inversely proportional
C) No relationship
D) Equal at all temperatures

22. Which of the following claims regarding how humidity affects air density is true?
A) Moist air is denser than dry air.
B) Humidity has no effect on air density.
C) Moist air is less dense than dry air.
D) Humidity always increases aircraft performance.

23. In relation to aircraft performance, what effect does an increase in weight have on the angle of attack?
A) Decreases the angle of attack
B) No change in the angle of attack
C) Increases the angle of attack
D) Weight does not influence the angle of attack

24. Which unit is equivalent to approximately 34 mb in measuring atmospheric pressure?
A) 1 °F
B) 1 "Hg
C) 1 PSI

D) 1 Pa

25. How do ground-based obstructions, like skyscrapers or tall trees, impact wind behavior for aviators?
A) They stabilize wind flow.
B) They have no effect on wind flow.
C) They produce gusts that vary in speed and direction.
D) They always cause a decrease in wind speed.

26. What is a significant danger for aircraft when encountering low-level wind shear?
A) Stable wind direction
B) Gradual changes in wind speed
C) Intense updrafts, downdrafts, and unexpected trajectory shifts
D) Constant headwind

27. In an unstable atmosphere, how do vertical currents react to minor disturbances?
A) They are suppressed.
B) They remain constant.
C) They amplify.
D) They reverse direction.

28. When does an atmospheric inversion occur?
A) When air temperature decreases with increasing altitude.
B) When air temperature remains constant with altitude.
C) When air temperature drops at ground level.
D) When air temperature rises with increasing altitude.

29.Which cloud type is particularly hazardous for pilots due to its turbulent nature and potential for severe weather conditions?
A) Cumulonimbus
B) Cirrus
C) Stratus
D) Nimbostratus

30. When navigating mountainous terrains with wind speeds exceeding 40 knots at mountain peak levels, pilots should:
A) Anticipate stable conditions.
B) Be less vigilant.
C) Anticipate turbulence.
D) Expect increased lift from windward sides.

31. Which of the following is NOT a performance drawback of an overweight aircraft?

A) Increased takeoff velocity
B) Augmented stalling velocity
C) Enhanced climb rate
D) Limited flight range

32. In steady, level flight, when does an aircraft achieve equilibrium?
A) When lift surpasses its weight
B) When lift diminishes below the weight
C) When the lift matches its weight
D) When the aircraft's weight exceeds its lift

33. What role does the center of gravity play in drones and other small crewless aircraft?
A) It determines the aircraft's top speed
B) It determines the maximum altitude
C) It significantly impacts stability
D) It decides the aircraft's fuel consumption

34. What transpires if the pressure center is located before the center of gravity?
A) The aircraft tends to pitch upwards
B) The aircraft remains neutral
C) The aircraft tends to pitch downwards
D) The aircraft rolls to one side

35. Which statement correctly explains load factors?
A) They describe the aircraft's maximum speed
B) They represent the force of gravity acting on an aircraft during turns
C) They reflect the intensity of force an object undergoes when accelerated
D) They detail the fuel consumption of an aircraft

36. In the context of aircraft weight and balance, what happens when the lift surpasses weight?
A) The vertical descent rate will amplify
B) The vertical speed will drop
C) The aircraft will hover in place
D) The vertical ascent rate will amplify

37. An aircraft's responsiveness to pilot directives, especially concerning its flight direction and posture, refers to which of the following?
A) Stability
B) Maneuverability
C) Controllability
D) Load Factor

38. What is the potential consequence of operating a drone designed for 3 Gs under a condition of 4 Gs?

A) The drone will ascend more quickly
B) The drone will gain additional endurance
C) The drone might be damaged or stall
D) The drone's battery life will extend

39. Which of the following effects of overloading an aircraft is NOT one?

A) Enhanced maneuverability
B) Diminished climb rate
C) Augmented stalling velocity
D) Extended takeoff distance

40. What does the force of weight in an aircraft counteract?

A) Thrust
B) Drag
C) Lift
D) Torque

41. When operating a drone near an airport without a control tower, why should a remote pilot monitor radio communication?

A) To provide continuous flight updates to manned aircraft.
B) To ensure their drone does not interfere with radio frequencies.
C) For heightened situational awareness and to identify nearby manned aircraft.
D) Because it is mandatory for all drone operations near any airport.

42. Which statement is true regarding the weight of an aircraft and its performance?

A) Overloading an aircraft enhances its performance.
B) The weight of an aircraft has no significant impact on its performance.
C) An aircraft's weight does not affect its takeoff or landing durations.
D) Overloading an aircraft result in diminished performance metrics.

43. What acts as the upward force on the wing of an aircraft?

A) Weight
B) Thrust
C) Drag
D) Lift

44. In aviation radio communication, how would the number 9 be pronounced to ensure clarity?

A) Nine

B) Nein
C) Nino
D) Niner

45. For manned aircraft approaching a non-towered airport, which typical announcement indicates their presence when they are 10 miles out?
A) "10 miles out, requesting permission to land."
B) "Approaching for landing, 10 miles away."
C) "10 miles east, ready for departure."
D) "[Airport Name] traffic, [Aircraft Call Sign], 10 miles [Direction], approaching for landing, [Airport Name] traffic."

46. In the context of hazardous attitudes for pilots, which attitude manifests as a tendency to disregard established guidelines or protocols?
A) Macho
B) Invulnerability
C) Anti-authority
D) Resignation

47. Which of the following is NOT one of the five prevalent hazardous attitudes that could compromise sound decision-making?
A) Overconfidence
B) Macho
C) Resignation
D) Impulsivity

48. Using the IMSAFE checklist, which factor concerns the consumption of drugs that might influence a pilot's cognitive abilities?
A) Illness
B) Medication
C) Alcohol
D) Emotion

49. According to the PAVE framework, which category would you evaluate to ensure the aircraft can manage the anticipated cargo and passenger weight?
A) Pilot's Capacity
B) Aircraft's Condition
C) Environment
D) External Influencers

50. Which hazardous attitude is characterized by a pilot feeling powerless or believing that their actions will not make a significant difference?

A) Macho
B) Invulnerability
C) Anti-authority
D) Resignation

51. When evaluating risks before a flight, what does the 'V' in the PAVE framework represent?
A) Vitality of the pilot
B) Variety of risks
C) Environment
D) Vision and clarity

52. In the context of evaluating risk, which external factor might compel pilots to overlook safety concerns due to an innate determination to achieve a goal?
A) Demonstrating piloting skills
B) Impressing an observer
C) The urgency to complete a mission
D) Admitting inexperience

53. Based on accessibility, which of the following is NOT a type of airfield?
A) Public Airfields
B) Controlled Airfields
C) Government/Military Airfields
D) Private Airfields

54. At uncontrolled airports, which frequency is allocated for communication to provide advisories while operating around the airport without an active control tower?
A) ATIS Frequency
B) Universal Integrated Community (UNICOM)
C) Common Traffic Advisory Frequency
D) FSS Frequency

55. What do VFR Urban Area Maps specifically target?
A) Military installations
B) Dense forest areas
C) Densely populated regions
D) International flight routes

56. What is the main purpose of the Automated Terminal Information Service (ATIS)?
A) Communicating with other pilots
B) Broadcasting the latest local weather and essential non-control details
C) Guiding pilots to their landing destination

D) Announcing new aviation regulations

57. What is the primary purpose of isogonic lines on aviation maps?
A) Indicating flight paths for commercial aircraft
B) Linking points with the same magnetic deviation
C) Showing regions with high radio frequency interference
D) Highlighting the areas prone to extreme weather conditions

58. On a sectional chart, what does a solid blue circle indicate?
A) Unpaved runway without a control tower
B) Paved runway with a control tower
C) Paved runway without a control tower
D) Detailed runway layout with a control tower

59. If an airport's runway length is depicted as "35" on a sectional chart, how long is the runway?
A) 3,500 feet
B) 350 feet
C) 35,000 feet
D) 3,500 meters

60. What does a magenta hashed outline denote on a sectional chart?
A) Restricted Area
B) Military Operation Area (MOA)
C) Warning Area
D) Prohibited Area

61. What symbol on a sectional chart indicates the presence of aviation fuel?
A) Small solid circle
B) Open wrench icon
C) Star inside a circle
D) Enclosed "W" symbol

62. What is the one thing that a sectional chart does NOT include?
A) Military Operation Areas (MOAs)
B) Controlled Firing Areas (CFAs)
C) Prohibited Areas
D) Alert Areas

63. What does a star inside a circle symbolize on a sectional chart?
A) Airport with customs services
B) Rotating airport beacon

C) Military Airport
D) Camping availability

64. On a sectional chart, which color represents areas of elevation typically below 2000 feet Mean Sea Level?
A) Green
B) Blue
C) Magenta
D) Light Brown

65. Which of the following areas requires right-hand turns for one or more runways and is indicated on a sectional chart?
A) Special Procedures
B) Right Traffic
C) Obstructions
D) Special Use Airspaces

66. On a sectional chart, what is the purpose of Maximum Elevation Figures (MEFs)?
A) To indicate the highest building in an area
B) To provide a quick reference for safe flying altitudes
C) To display the chart's name and edition number
D) To depict the highest mountain peak

67. What is represented by a parachute symbol on a sectional chart?
A) Restricted Area
B) Military Operation Area
C) Seaplane Base
D) Parachute Jumping Area

68. Which symbol on a sectional chart indicates a heliport?
A) H inside a circle
B) Tent symbol
C) Anchor symbol
D) Enclosed "W"

69. What does a solid magenta circle denote on a sectional chart?
A) Paved runway without a control tower
B) Paved runway with a control tower
C) Unpaved runway without a control tower
D) Detailed runway layout with a control tower

70. What type of landmark can be a bridge, dam, or stadium on a sectional chart?

A) Natural Landmark
B) Man-made Landmark
C) Airport Landmark
D) Special Procedures Landmark

71. On a sectional chart, which areas are outlined with blue hashed lines and labeled with an "R" followed by a number?
A) Restricted Areas
B) Military Operation Areas
C) Warning Areas
D) Prohibited Areas

72. Where on the sectional chart can pilots find information about the chart's name, its current edition, and the scale?
A) In the center of the chart
B) Alongside visual checkpoints
C) In the margins of the chart
D) Next to each airport symbol

Practice Test - Answers

1. B

Class B airspace has a unique configuration comprising a surface region and multiple layers, which can sometimes be visualized as inverted wedding cakes. This unique structure ensures the smooth flow of traffic in and out of the nation's busiest airports.

2. D

Class E airspace represents the controlled regions not categorized under Class B, C, or D. With an usual beginning point of 1,200 feet above ground level (AGL), Class E airspace makes up a sizable section of the airspace throughout the United States. Class E airspace rises to 18,000 feet MSL but does not include that height.

3. C

Class G airspace, often referred to as uncontrolled airspace, starts from the ground level and extends upward until it meets the base of the overlying Class E airspace. Remote pilots are not needed to get air traffic control clearance when operating in Class G airspace, in contrast to the controlled classes of airspace.

4. D

Controlled Firing Areas (CFAs) host operations that could pose risks to non-participating aircraft if not managed under strict conditions. The distinctive feature of CFAs is that activities within them will halt when a non-participating aircraft is detected to be approaching. They are not marked on charts because their operations do not require non-participating aircraft to change their routes.

5. A

Warning Areas are symbolized on charts with a "W" prefix, followed by an identifier. They have characteristics like restricted areas but are distinct in that they start from 3 NM beyond the U.S. coastline and may contain activities that are potentially hazardous to non-participating aircraft.

6. D

Prohibited zones are specific areas of airspace where aircraft movement is forbidden. These areas are established due to security reasons or other factors critical to national interests. Details about these areas, such as the vicinity around Area 51 in Nevada, are published in the Federal Register and can be seen on aeronautical charts.

7. B

The Local Airport Advisory (LAA) is a service presented by Flight Service stations positioned at the airport. It operates via a specific ground-to-air channel and offers insights about the local airport, automated weather updates, and uninterrupted data from systems like the Automated Surface Observing System or the Automated Weather Observing Station.

8. B

Military Training Routes that remain entirely below 1,500 feet AGL use a four-digit code (like IF3012, VF3013). Those with parts rising above 1,500 feet AGL use a three-digit system (like IF312, VF313).

9. B

Temporary Flight Restrictions are established for several reasons including shielding individuals and assets, ensuring the safety of disaster relief aircraft, mitigating overcrowded skies above remarkable events, safeguarding national disaster zones for compassionate reasons, ensuring the safety of high-profile personalities, and facilitating secure conditions for space agency missions.

10. B

Notices to Airmen can be accessed in printed formats via subscriptions from the Document Oversight Office or digitally through platforms like PilotWeb, which offers the latest NOTAM insights. Various online portals also offer localized airport NOTAM details.

11. A

The METAR reports wind details as 15018G24KT, which indicates the wind is coming from 150° at 18 knots and gusting up to 24 knots.

12. A

In a METAR report, "AUTO" indicates that the report is from an automated source.

13. B

The METAR reports visibility as "2SM," which stands for 2 statute miles.

14. A

TAF KJFK 121400Z 1214/1314 indicates the report was issued on the 12th day of the month at 1400Z and covers a 24-hour forecast period.

15. C

"TSRA" is an abbreviation used in METAR and TAF reports to denote a thunderstorm with rain.

16. B

In AIRMETs, the code "TANGO" signifies turbulence.

17. B

Convective SIGMETs are primarily issued for severe thunderstorms that display ground winds exceeding 50 knots, large hail, or tornadoes.

18. A

The description provides the boundary or perimeter of the affected area by listing various waypoints, indicating the turbulence affects the entire region defined by these points.

19. C

A Terminal Aerodrome Forecast (TAF) is designed to detail the atmospheric conditions within a five-statute mile radius of an airport.

20. B

In a Convective SIGMET, "PROB30" describes the 30% likelihood of thunderstorms and rain occurring within the forecast period.

21.B

As the temperature rises, the density of air typically decreases. Conversely, as the temperature drops, air density increases. As a result, with constant pressure, air density and temperature are inversely proportional.

22.C

Water vapor is lighter than air. Thus, when the air's moisture content increases, its density drops, making moist air less dense than dry air.

23. C

An increase in weight necessitates the plane to maintain a steeper angle of attack to sustain a specific altitude and speed.

24. B

1 "Hg (inches of mercury) is roughly equivalent to 34 mb (millibars) in atmospheric pressure measurements.

25. C

Ground-based obstructions can disrupt the steady flow of the wind, producing gusts that can change rapidly in speed and direction.

26. C

Wind shear refers to rapid variations in wind direction or speed over short distances. Encountering such changes can result in strong updrafts, downdrafts, and unexpected shifts in the aircraft's path, especially dangerous at low altitudes.

27. C

In an unstable atmosphere, minimal vertical currents can escalate, resulting in turbulence and potential convective phenomena.

28.D

An inversion occurs when, contrary to the usual pattern, the air temperature rises as altitude increases.

29.A

Cumulonimbus clouds result from rising air currents, making them turbulent and posing significant flight risks. These clouds can bring about large hail, strong lightning, tornadoes, and heavy rainfall.

30. C

Wind speeds surpassing 25 knots at mountain peak levels suggest potential turbulence, and speeds over 40 knots require heightened caution. Layered cloud formations indicate stable air conditions, but the presence of standing lenticular or rotor clouds can be a sign of mountain waves, leading to turbulence.

31. C

An overweight aircraft would typically have a diminished climb rate, not an enhanced one. The weight would make it more challenging for the aircraft to ascend rapidly.

32. C

In steady, level flight, equilibrium is achieved when the lift produced by the aircraft equals its weight, preventing any upward or downward acceleration.

33. C

The center of gravity is a crucial factor in determining the stability of drones and other small crewless aircraft. A shift or imbalance in the center of gravity can affect the flight characteristics and stability.

34. C

When the center of gravity is ahead of the center of pressure, the aircraft will have a nose-down tendency, causing it to pitch downwards.

35. C

Load factors are expressed in gravitational acceleration units (Gs) and represent the stress on an aircraft's structure, reflecting the force experienced when an object (like an aircraft) is accelerated.

36. D

If lift surpasses weight, the aircraft will experience a positive vertical acceleration or an amplified ascent rate.

37. C

Controllability pertains to how the aircraft reacts to pilot inputs and its responsiveness, especially in terms of its flight direction and posture.

38. C

Subjecting a drone to a load factor higher than its design can lead to structural damage or cause the drone to stall, especially if operated at speeds perceived as safe under normal conditions.

39. A

Overburdening an aircraft typically results in reduced performance, including hampered agility. It does not enhance maneuverability.

40. C

Weight is the force exerted by gravity pulling the aircraft towards the Earth's core. Lift acts as the upward force on the wing, counteracting the aircraft's weight and allowing it to stay in the air.

41. C

While it is not mandatory for drone pilots to communicate with manned aircraft near non-towered airports, understanding the traffic patterns and radio procedures can help enhance safety. Monitoring these communications provides a better understanding of the positions and intentions of manned aircraft, enabling remote pilots to ensure safe operations and prevent potential collisions.

42.D

An overloaded aircraft displays compromised flight behaviors. Critical performance drawbacks of an overweight aircraft include an increased takeoff velocity, extended takeoff distance, diminished climb rate, limited maximum altitude, reduced flight range, and prolonged landing duration, among others.

43. D

Lift acts as the upward force on the wing, perpendicular to both the oncoming airflow and the aircraft's side axis. It counteracts the aircraft's weight, and in steady, level flight, the aircraft achieves equilibrium when the lift matches its weight.

44. D

Numbers should be articulated clearly in aviation communication to prevent confusion. The number 9 is pronounced as "niner" to ensure it does not get confused with similar sounding words or terms.

45. D

Explanation:

As manned aircraft approach a non-towered airport, they typically announce their presence and intentions when they are 10 miles out, indicating their relative position (north, south, east, or west). This communication helps other aircraft in the vicinity understand their position and trajectory.

46. C

The anti-authority attitude is characterized by a disregard for established guidelines or rules. This attitude can lead a pilot to ignore essential protocols or instructions, believing they know better or that the rules do not apply to them.

47. A

The five dangerous attitudes that have been found to be most common are contempt for authority, impulsivity, invulnerability, machismo, and resignation. Overconfidence, while a potential issue, is not explicitly listed among these five.

48. B

Illness, Medication, Stress, Alcohol, Fatigue, and Emotion make up the IMSAFE checklist. The 'Medication' component focuses on any drugs that a pilot might have consumed that could impair cognitive abilities or induce drowsiness.

49. B

The PAVE framework divides flight risks into Pilot's capacity (P), Aircraft's condition (A), Environment (V), and External influencers (E). The aircraft's ability to handle specific weights would fall under the 'Aircraft's Condition' category.

50. D

The resignation attitude makes a pilot feel powerless or adopt a fatalistic view, believing whatever is destined to happen will happen, regardless of their actions or decisions.

51. C

In the PAVE framework, 'V' stands for Environment. This category involves evaluating weather conditions, landscape, and airspace, among other environmental factors.

52. C

External influencers can sometimes overshadow safety concerns. An innate determination to complete a mission or achieve set goals can serve as a significant external factor pushing pilots to potentially risky behaviors.

53. B

The provided information categorizes airfields based on accessibility into Public Airfields, Government/Military Airfields, and Private Airfields. "Controlled Airfields" is a classification based on the presence or absence of a control tower, not accessibility.

54. C

The Common Traffic Advisory Frequency is allocated for communication to provide advisories while operating around an airport without an active control tower. This frequency can be tied to a UNICOM, MULTICOM, FSS, or a tower as highlighted in relevant aviation documents.

55. C

VFR Urban Area Maps specifically target densely populated regions, providing a more detailed view of city landscapes. This granularity is beneficial for drone operators and other pilots operating in urban settings, making them aware of obstacles, notable buildings, and other urban-specific navigation aids.

56. B

The Automated Terminal Information Service (ATIS) continuously broadcasts the latest local weather and other essential non-control details on a designated frequency. It provides information such as weather conditions, active runways, and specific air traffic control procedures.

57. B

Isogonic lines connect locations with the same magnetic deflection on aviation charts. These lines show how true north (TN) and magnetic north differ from one another (MN). The precise deviation at numerous locations has been charted, and these isogonic lines display these deviations on the map. The direction and magnitude of deviation might undergo slight changes periodically, influenced by distinct geological factors that affect the magnetic forces in those regions.

58. B

A solid blue circle indicates a paved runway with a control tower.

59. A

The number represents the runway length in hundreds of feet. Thus, "35" would indicate a length of 3,500 feet.

60. B

A magenta hashed outline represents a Military Operation Area (MOA).

61. A

A small solid circle denotes the availability of aviation fuel.

62. B

Controlled Firing Areas (CFAs) are not depicted on sectional charts because the activities pose no hazard to aircraft.

63. B

A star inside a circle symbolizes the presence of a rotating airport beacon, usually operational from sunset to sunrise.

64. A

Green represents areas of lower elevation, typically below 2000 feet Mean Sea Level.

65. B

If an airport has a traffic pattern that requires right-hand turns for one or more runways, it will be indicated with an "R" and the runway number.

66. B

Maximum Elevation Figures (MEFs) provide a quick reference for safe flying altitudes, ensuring clearance of all obstacles and terrain in each quadrant.

67. D

A parachute symbol denotes an area where parachute jumping activity occurs.

68. A

A circle with an "H" in the middle represents a heliport.

69. A

A solid magenta circle indicates a paved runway without a control tower.

70. B

Explanation: Man-made landmarks encompass structures like bridges, dams, tall buildings, factories, stadiums, and distinct road intersections.

71. A

Explanation: Restricted Areas are outlined with blue hashed lines and labeled with an "R" followed by a number.

72. C

Explanation: Marginal information on sectional charts, located in the margins, provides essential details such as the chart's name, its current edition number or date, and the scale.

Appendix

Glossary

Aeronautical Chart: A map used for flight navigation that displays information such as topography, airspace classes, and airport data.

AGL (Above Ground Level): A measurement of altitude concerning the ground beneath the aircraft.

Air Traffic Control (ATC): A service that manages aircraft on the ground and in the air to ensure safe operations.

Airspace: The area of the atmosphere that is located above a certain geographical mass, often a country or continent

Automated Terminal Information Service (ATIS): An ongoing log of the local weather and other non-control information

BVLOS (Beyond Visual Line of Sight): Refers to drone operations where the remote pilot or observer cannot see the drone without visual aids, such as binoculars or telescopes.

CTAF (Common Traffic Advisory Frequency): A channel that pilots use to signal their intentions when flying to or from an airport without a functioning control tower.

Drone: Another term for an unmanned aerial vehicle (UAV) or unmanned aircraft system (UAS), typically referring to a multirotor or quadcopter used for various applications.

FAA (Federal Aviation Administration): The branch of American government in charge of aviation security.

FPV (First Person View): A method of piloting where the operator controls the drone as if they are onboard, typically using a live video feed.

Geo-fencing: A feature in a software program that uses the global positioning system (GPS) to define geographical boundaries, often used in drones to prevent them from flying into restricted areas.

Latitude and Longitude: Reference lines on the Earth's surface used for navigation and location identification.

Magnetic North (MN): The opposite of true north as determined by the Earth's magnetic field and the direction that a compass needle points

NOTAM (Notice to Airmen): A notice containing information that could affect a pilot's decision to enter a particular area or change flight paths, such as hazards or temporary flight restrictions.

Part 107: The common name for the FAA's regulations governing the commercial use of drones or small unmanned aerial systems (sUAS) in the U.S.

Payload: The carrying capacity of an aircraft or drone, usually referring to something other than the drone's standard equipment. This could be cameras, sensors, or other equipment.

Remote Pilot in Command (RPIC): The individual who is in charge of the drone flight's operation and safety in the end analysis

Sectional Chart: A specific kind of aeronautical chart used for slow- to medium-speed aircraft's visual navigation.

sUAS (Small Unmanned Aircraft Systems): Drones or unmanned aircraft with its payload and equipment weighing less than 55 pounds

Telemetry: A method of automated communication used to transfer data to receiving equipment for monitoring from devices placed at distant or inaccessible locations.

True North (TN): The direction that is pointed in the direction of the North Pole.

Unmanned Aircraft System (UAS): An unmanned aircraft and the tools required to operate it, such as navigational aids and communication systems.

Variation: The angle between true north and magnetic north, as defined by the magnetic field in the area.

Waypoint Navigation: A sequence of geographical positions, used for navigation, where a drone is guided from point to point.

Yaw, Pitch, and Roll: The three axes in which a drone or any aircraft moves. Yaw is the left and right movement, pitch is the up and down movement, and roll is the tilting movement from side to side.

Phonetic Alphabet

The International Civil Aviation Organization (ICAO) phonetic alphabet, often referred to as the NATO phonetic alphabet, is used internationally for radio communications. Here it is:

A - Alpha
B - Bravo
C - Charlie
D - Delta
E - Echo
F - Foxtrot
G - Golf
H - Hotel
I - India
J - Juliett
K - Kilo
L - Lima
M - Mike
N - November
O - Oscar
P - Papa
Q - Quebec
R - Romeo
S - Sierra
T - Tango
U - Uniform
V - Victor
W - Whiskey
X - X-ray
Y - Yankee
Z - Zulu

Additionally, numbers are typically pronounced in a specific way in radio communications to avoid confusion:

0 - Zero
1 - One
2 - Two
3 - Three
4 - Four
5 - Five
6 - Six

7 - Seven
8 - Eight
9 - Niner

Answers

METARs

1. METAR KJFK 221152Z 28016KT 10SM FEW035 SCT070 BKN200 03/M03 A3005 RMK AO2 SLP174 T00281028 10028 21044 53015

- METAR: Routine standard meteorological report.
- KJFK: The John F. Kennedy International Airport in New York is the subject of the report.
- 221152Z: The report was recorded on the 22nd day at 1152Z (UTC).
- 28016KT: Winds are coming from 280° at 16 knots.
- 10SM: Visibility is 10 statute miles.
- FEW035: There are a few clouds at 3,500 feet AGL.
- SCT070: Scattered clouds at 7,000 feet AGL.
- BKN200: Broken cloud layer at 20,000 feet AGL.
- 03/M03: The temperature is 3°C, and the dew point is -3°C.
- A3005: The barometric pressure is 30.05 inches of mercury.
- RMK AO2 SLP174 T00281028 10028 21044 53015: This is the remark section indicating various additional details.

2. METAR EGLL 221150Z 07012KT 040V100 9000 -DZ FEW008 SCT015 BKN040 05/03 Q1018 NOSIG

- METAR: Routine standard meteorological report.
- EGLL: The report is for London Heathrow Airport.
- 221150Z: The report was recorded on the 22nd day at 1150Z.
- 07012KT 040V100: Winds are from 70° at 12 knots but varying between 40° and 100°.
- 9000: Visibility is 9,000 meters.
- -DZ: Light drizzle.
- FEW008: There are a few clouds at 800 feet AGL.
- SCT015: Scattered clouds at 1,500 feet AGL.
- BKN040: Broken cloud layer at 4,000 feet AGL.
- 05/03: Temperature is 5°C, and the dew point is 3°C.
- Q1018: Barometric pressure is 1018 hPa.
- NOSIG: No significant weather change expected soon.

3. METAR YSSY 221200Z 04008KT 9999 FEW030 SCT048 22/16 Q1016

- METAR: Routine standard meteorological report.
- YSSY: The report is for Sydney Kingsford Smith Airport in Australia.
- 221200Z: The report was recorded on the 22nd day at 1200Z.

- 04008KT: Winds are from 40° at 8 knots.
- 9999: Visibility is 10,000 meters or more.
- FEW030: There are a few clouds at 3,000 feet AGL.
- SCT048: Scattered clouds at 4,800 feet AGL.
- 22/16: Temperature is 22°C, and the dew point is 16°C.
- Q1016: Barometric pressure is 1016 hPa.

TAFs

1. TAF KLAX 221400Z 2214/2314 27010KT P6SM SCT020 BKN050 FM222000 25015KT P6SM SCT025 SCT200 FM230300 28008KT 6SM BR SCT015 BKN030 PROB30 2306/2310 4SM BR BKN015

- Routine forecast for Los Angeles International Airport (KLAX) issued on the 22nd day of the month at 1400Z.
- The forecast is valid from 1400Z on the 22nd to 1400Z on the 23rd.
- Initially, the winds are coming from 270° at 10 knots. Visibility is beyond 6 statute miles (P6SM) with scattered clouds at 2000 feet and broken clouds at 5000 feet.
- From 2000Z, winds shift to come from 250° at 15 knots, still with visibility beyond 6 statute miles. At an altitude of 20,000 feet and 2500 feet, respectively, there will be scattered clouds.
- From 0300Z on the 23rd, winds are from 280° at 8 knots. Visibility reduces to 6 statute miles with mist (BR). There will be scattered clouds at 1500 feet and broken clouds at 3000 feet.
- There is a 30% probability between 0600Z and 1000Z on the 23rd that visibility will be 4 statute miles due to mist with broken clouds at 1500 feet.

2. TAF EGLL 221400Z 2214/2314 06008KT 9999 SCT030 BKN080 TEMPO 2218/2222 4000 -RA BKN025 PROB40 2300/2304 3SM RA BKN015 FM230500 08012KT P6SM BKN020 OVC040

- Routine forecast for London Heathrow Airport (EGLL) issued on the 22nd at 1400Z.
- Valid from 1400Z on the 22nd to 1400Z on the 23rd.
- Winds are from 060° at 8 knots with visibility at 10 kilometers (which is equivalent to 6 statute miles). At 8000 feet, there are broken clouds and dispersed clouds, respectively.
- Temporarily between 1800Z and 2200Z on the 22nd, visibility drops to 4000 meters with light rain and broken clouds at 2500 feet.
- There is a 40% chance between 0000Z and 0400Z on the 23rd that visibility will be 3 statute miles due to rain with broken clouds at 1500 feet.
- From 0500Z on the 23rd, winds will be from 080° at 12 knots with visibility beyond 6 statute miles. Broken clouds will be at 2000 feet and overcast at 4000 feet.

3. TAF YSSY 221400Z 2214/2314 04008KT 9999 FEW030 SCT060 FM221800 02010KT 8000 -SHRA SCT020 BKN040 TEMPO 2220/2224 5000 TSRA BKN025CB FM230400 05006KT P6SM SCT030 SCT060

- Routine forecast for Sydney Airport (YSSY) issued on the 22nd at 1400Z.
- Valid from 1400Z on the 22nd to 1400Z on the 23rd.
- Winds are blowing at 8 knots and have a 10 km visibility. At 3000 feet, there are a few clouds, and at 6000 feet, there are sporadic clouds.
- From 1800Z on the 22nd, winds shift to come from 020° at 10 knots. Visibility reduces to 8000 meters with light rain showers. At 2000 feet, scattered clouds will be present, and at 4000 feet, broken clouds.
- Temporarily between 2000Z and 2400Z on the 22nd, visibility further drops to 5000 meters due to thunderstorms with rain. Broken cumulonimbus clouds will be at 2500 feet.
- From 0400Z on the 23rd, winds will be from 050° at 6 knots with visibility beyond 6 statute miles. Scattered clouds at both 3000 feet and 6000 feet.

WSTs

1. WST 17E WT 231850

This is a Convective SIGMET issued for the Eastern region (E) and is the 17th report of the month. It was issued on the 23rd day at 1850Z. This is the second update (UPDT 2) for the AIRMET Sierra, which warns of Instrument Flight Rules (IFR) conditions and mountain obstructions. The warning is valid until the 24th day at 0200Z. IFR conditions are present in portions of California (CA) and Nevada (NV). The boundary of the affected area is defined by the waypoints or navigation aids starting from 40NNE MOD, going to 30SSE BTY, then to 40W LAS, to 50N RZS, and back to 40NNE MOD. Within this area, the ceiling is below 010 (1,000 feet) and visibility is below 3 statute miles due to precipitation and mist. These conditions are expected to end between 0000Z and 0300Z.

2. WST 09W WT 231720

This Convective SIGMET is issued for the Western region (W) and is the 9th report of the month. The report was given on the 23rd day at 1720Z. This is the first update (UPDT 1) for the AIRMET Zulu, which warns of icing and the freezing level. The warning is in effect until the 24th day at midnight (0000Z). The areas affected by the icing conditions are parts of Oregon (OR) and Washington (WA). The boundary of the area begins at 20W HUH, goes to 40SSW PDT, then to 50SE DSD, 30N OLM, and finally back to 20W HUH. Within this area, moderate icing is expected between Flight Level 18,000 feet (FL180) and Flight Level 26,000 feet (FL260). These conditions are anticipated to continue beyond midnight (00Z) and will persist until 0600Z.

3. WST 22S WT 232115

This Convective SIGMET was issued for the Southern region (S) and is the 22nd report of the month. It was dispatched on the 23rd day at 2115Z. This is the fourth update (UPDT 4) for the

AIRMET Sierra, warning of IFR conditions and mountain obstructions. The warning is valid until the 24th day at 0400Z. The areas under the influence of IFR conditions include parts of Colorado (CO), Utah (UT), and Arizona (AZ). The affected area's boundary is defined starting from 30NNW DEN, moving to 20SE MTJ, then to 40ESE PHX, to 50SSW BCE, and returning to 30NNW DEN. Within this boundary, the ceiling is below 005 (500 feet) and visibility is less than 2 statute miles due to mist and fog. These conditions are predicted to end between 0200Z and 0500Z.

Digital Bonuses

Dear Reader,
As a token of our appreciation for your commitment and interest, we're thrilled to offer you an exclusive bonus! Gain access to a special set of flashcards packed with sectional charts exercises designed to boost your understanding and proficiency.

Here's how to get your bonus:

1. Locate the QR Code: At the bottom of this page, you'll find a unique QR code.
2. Scan the Code: Use your smartphone or tablet's camera to scan the QR code. Most devices will recognize and process the QR code automatically, opening a link or prompt.
3. Access the Flashcards: Once you've scanned the code, you'll be directed to our exclusive flashcards platform. Dive in and start practicing right away!

Remember, these flashcards are a bonus feature and won't be available in any other format. Make sure you grab this opportunity to enrich your learning experience.

Thank you for being an active participant in your educational journey. We're excited for you to explore these exercises and elevate your skills even further.

Happy Studying!

Made in the USA
Las Vegas, NV
18 October 2024